THE PERFECT THEORY

"Einstein's beautiful theory is almost a century old, and its ramifications have stimulated a crescendo of discovery ever since. It is now, more than ever, one of the liveliest frontiers of science. Pedro G. Ferreira describes, accessibly and non-technically, how the key breakthroughs have been made, and the personalities who made them. Even readers with zero scientific background will enjoy this finely written survey of one of the greatest of recent scientific endeavours, and get a real feel for the social and human aspects of science."
— Martin Rees, Lord Rees of Ludlow, Astronomer Royal

"*The Perfect Theory* is a rollicking good read. We watch as Einstein's brilliant successors struggle and squabble about everything from black holes to quantum gravity. With crisp explanations and narrative flair, Ferreira offers us a fun, fresh take on a magnificent part of modern science."
— Steven Strogatz, Schurman Professor of Applied Mathematics, Cornell University, author of *The Joy of X*

"Einstein's general theory of relativity was the greatest of his many contributions to physics, but surprisingly little has been written about how the subject blossomed after his death, with profound implications for current cosmology and astrophysics. Pedro G. Ferreira provides an enthralling account of the ideas and personalities of those involved."
— Sir Roger Penrose, Rouse Ball Professor of Mathematics Emeritus, University of Oxford, author of *The Road to Reality*

"You couldn't ask for a better guide to the outer reaches of the universe and the inner workings of the minds of those who've navigated it."
— Marcus du Sautoy, Simonyi Professor for the Public Understanding of Science, University of Oxford, author of *The Music of the Primes*

"Pedro G. Ferreira portrays a community ensnared by a single great idea. His telling is incredibly thorough while still beautiful, deeply considered and affecting. With vivid detail, he brings to life the awesome story of one of humanity's greatest achievements."
— Janna Levin, Professor of Physics and Astronomy, Barnard College/ Columbia University, author of *How the Universe Got Its Spots*

"Einstein's general relativity is a theory of unrivaled elegance and simplicity. But the history of general relativity is messy, unpredictable, and occasionally dramatic. Pedro G. Ferreira is an expert guide to the twists and turns scientists have gone through in a quest to understand space and time."
— Sean Carroll, theoretical physicist, author of *The Particle at the End of the Universe*

"This is a fascinating introduction to our present understanding of space, time, and gravity, and to the confusion about how to go about finding a still better theory. Ferreira tells the story without equations or graphs, just well-chosen words about the science and how it grew. I particularly recommend the sketches of scientists in all their curious variety of character traits."
— P. James Peebles, Albert Einstein Professor of Science Emeritus, Princeton University

THE PERFECT THEORY

*A Century of Geniuses and the
Battle over General Relativity*

PEDRO G. FERREIRA

Little, Brown

LITTLE, BROWN

First published in Great Britain in 2014 by Little, Brown

Copyright © Pedro G. Ferreira 2014

The right of Pedro Ferreira to be identified as author
of this work has been asserted by him in accordance with
the Copyright, Designs and Patents Act 1988.

A CIP catalogue record for this book
is available from the British Library.

Hardback ISBN 978-1-4087-0310-6
C format ISBN 978-1-4087-0430-1

Printed and bound in Great Britain by
Clays Ltd, St Ives plc

Papers used by Little, Brown are from well-managed forests
and other responsible sources.

MIX
Paper from
responsible sources
FSC
www.fsc.org   FSC® C104740

Little, Brown
An imprint of
Little, Brown Book Group
100 Victoria Embankment
London EC4Y 0DY

An Hachette UK Company
www.hachette.co.uk

www.littlebrown.co.uk

To Gisa, Bruno, and Mia

Contents

PROLOGUE   ix

1. IF A PERSON FALLS FREELY   1
2. THE MOST VALUABLE DISCOVERY   12
3. CORRECT MATHEMATICS, ABOMINABLE PHYSICS   28
4. COLLAPSING STARS   47
5. COMPLETELY CUCKOO   66
6. RADIO DAYS   85
7. WHEELERISMS   100
8. SINGULARITIES   118
9. UNIFICATION WOES   137
10. SEEING GRAVITY   152
11. THE DARK UNIVERSE   173
12. THE END OF SPACETIME   193
13. A SPECTACULAR EXTRAPOLATION   209
14. SOMETHING IS GOING TO HAPPEN   223

ACKNOWLEDGMENTS   237
NOTES   239
BIBLIOGRAPHY   258
INDEX   270

Prologue

W HEN ARTHUR EDDINGTON stood up at a joint meeting of the Royal Society and Royal Astronomical Society
on November 6, 1919, his announcement quietly upended
the reigning paradigm of gravitational physics. In a solemn mono-
tone, the Cambridge astronomer described his trip to the small, lush
island of Príncipe off the west coast of Africa, where he had set up a
telescope and taken photographs of a total eclipse of the sun, being
particularly careful to capture a faint cluster of stars scattered behind
it. By measuring the positions of those stars, Eddington had found
that the theory of gravity invented by British science's patron saint,
Isaac Newton, a theory that had been accepted as truth for over two
centuries, was wrong. In its place, he claimed, belonged a new, correct
theory proposed by Albert Einstein, known as the "general theory of
relativity."

At the time, Einstein's theory was already known as much for its
potential to explain the universe as it was for its incredible difficulty.
After the ceremony, as the audience and speakers milled around, ready
to escape into the London evening, a Polish physicist named Ludwik
Silberstein ambled over to Eddington. Silberstein had already written
a book about Einstein's more restricted "special theory of relativity"
and had followed Eddington's presentation with interest. Now he pro-
nounced, "Professor Eddington, you must be one of the three persons
in the world who understand general relativity." When Eddington was

slow to respond, he added, "Don't be modest, Eddington." Eddington looked at him firmly and said, "On the contrary, I am trying to think who the third person is."

By the time I first discovered Einstein's general theory of relativity, Silberstein's count could probably be adjusted upward. It was the early 1980s, and I saw Carl Sagan on the television series *Cosmos* talking about how space and time can shrink or stretch. I immediately asked my dad to explain the theory. All he could tell me was that it is very, very difficult. "Hardly anyone understands general relativity," he said. I was not so easily deterred. There was something deeply appealing about this bizarre theory, with its warped grids of spacetime wrapping around deep, desolate throats of nothingness. I could see general relativity at work on old episodes of *Star Trek* when the starship *Enterprise* was kicked back in time by a "black star" or when James T. Kirk floundered around between different dimensions of spacetime. Could it really be so hard to understand?

A few years later I went to university in Lisbon, where I studied engineering in a monolithic building of stone, iron, and glass, a perfect example of the Fascist architecture of the Salazar regime. The setting was apt for the endless lectures where we were taught useful things: how to build computers, bridges, and machines. A few of us escaped the drudgery by reading about modern physics in our spare time. We all wanted to be Albert Einstein. Every now and then some of his ideas would appear in our lectures. We learned how energy is related to mass and how light is actually made of particles. When it was time to study electromagnetic waves, we were introduced to Einstein's special theory of relativity. He had come up with it in 1905, at the tender young age of twenty-six, just a few years older than us. One of our more enlightened lecturers told us to read Einstein's original papers. They were little gems of concision and clarity compared to the tedious exercises we were being set. But general relativity, Einstein's grand theory of spacetime, was not part of the menu.

At some point I decided to teach myself general relativity. I scoured the library at my university and found a mesmerizing collection of monographs and textbooks by some of the greatest physicists and mathematicians of the twentieth century. There was Arthur Eddington, the Astronomer Royal from Cambridge; Hermann Weyl, the geometer from Göttingen; Erwin Schrödinger and Wolfgang Pauli, both fathers of quantum physics — all with their own take on how Einstein's theory should be taught. One tome looked like a big black phone book, running more than a thousand pages, with flourishes and comments from a trio of American relativists. Another one, written by the quantum physicist Paul Dirac, barely made it to a sleek and spare seventy pages. I felt that I had entered a completely new universe of ideas populated by the most fascinating characters.

Understanding their ideas wasn't easy. I had to teach myself to think in a completely new way, relying on what initially seemed like elusive geometry and abstruse mathematics. Decoding Einstein's theory required mastering a foreign mathematical language. Little did I know that Einstein himself had done the same as he tried to figure out his own theory. Once I learned the vocabulary and grammar, I was blown away by what I could do. And so began my lifelong love affair with general relativity.

It sounds like the ultimate overstatement, but I can't resist it: the reward for harnessing Albert Einstein's general theory of relativity is nothing less than the key to understanding the history of the universe, the origin of time, and the evolution of all the stars and galaxies in the cosmos. General relativity can tell us about what lies at the farthest reaches of the universe and explain how that knowledge affects our existence here and now. Einstein's theory also sheds light on the smallest scales of existence, where the highest-energy particles can come into being out of nothing. It can explain how the fabric of reality, space, and time emerges to become the backbone of nature.

What I learned during those months of intense study is that general relativity brings space and time alive. Space is no longer just a place where things exist, nor is time a ticking clock keeping tabs on things. According to Einstein, space and time are intertwined in a cosmic dance as they respond to every single speck of stuff imaginable, from particles to galaxies, weaving themselves into elaborate patterns that can lead to the most bizarre effects. And from the moment he first proposed his theory, it has been used to explore the natural world, revealing the universe as a dynamic place, expanding at breakneck speed, filled with black holes, devastating punctures of space and time, and grand waves of energy, each carrying almost as much energy as a whole galaxy. General relativity has let us reach further than we ever imagined.

There was something else that struck me when I first learned general relativity. Although Einstein took just under a decade to develop it, it has remained unchanged ever since. For almost a century, it has been considered by many to be the perfect theory, a source of profound admiration to anyone who has had the privilege of coming across it. General relativity has become iconic for its resilience, as a centerpiece of modern thought and as a colossal cultural achievement along the lines of the Sistine Chapel, the Bach cello suites, or an Antonioni film. General relativity can be encapsulated succinctly in a set of equations and rules that are easy to summarize and write down. And they are not just beautiful — they also say something about the real world. They have been used to make predictions about the universe that have since been proved by observation, and there is a firm belief that buried in general relativity there are more deep secrets about the universe that remain to be exposed. What more could I want?

For almost twenty-five years, general relativity has been part of my daily life. It has been at the heart of much of my research and under-pinned so much of what my collaborators and I are trying to under-stand. My first experience with Einstein's theory was far from unique; I have come across people from all over the world who have been hooked by Einstein's theory and have devoted their lives to uncovering its mysteries. And I really do mean all over the world. From Kinshasa to Kraków, and from Canterbury to Santiago, I am regularly sent scientific papers whose authors are trying to find new solutions or even possible changes to general relativity. Einstein's theory may be difficult to grasp but it is also democratic; its very difficulty and intractability mean that much remains to be done before all its implications are exposed. There is opportunity here for anyone with a pen and paper and stamina.

I have often heard PhD supervisors telling their students not to work on general relativity for fear of becoming unemployable. To many it is far too esoteric. Devoting one's life to general relativity is definitely a labor of love, an almost irresponsible calling. But once you have been bitten by the bug, it can be all but impossible to leave relativity behind. Recently I met one of the leading lights in modeling climate change. He is a real pioneer in the field, a fellow of the Royal Society, an expert in making predictions of weather and climate in what is still a fiendishly difficult area of research. He hasn't always done this for a living. In fact, as a young man in the 1970s, he studied general relativity. That was almost forty years ago, yet, when we first met, he told me with a wry smile, "I am, in fact, a relativist."

A friend of mine left academia quite a while back, having worked on Einstein's theory for almost twenty years. He now works for a software company, developing and installing mechanisms for storing large amounts of data. His week is spent flying all over the world to set up these highly complex and expensive systems in banks, corporations, and government offices. Yet when we meet, he always wants to quiz me about Einstein's theory, or share with me his latest thoughts on general relativity. He can't shake it.

One thing about general relativity that has always puzzled me is how, despite being around for almost a century, it continues to yield new results. I would have thought that, given the phenomenal brain power that has been devoted to it, the theory would have been done and dusted decades ago. The theory might be difficult, but surely there is a limit to how much it can give us? Aren't black holes and an expanding universe more than enough? But as I've continued to grapple with the ideas that come out of Einstein's theory and met many of the brilliant minds that have worked on it, it has dawned on me that the story of general relativity is a fascinating and magnificent narrative, maybe as complicated as the theory itself. The key to understanding why the theory is, well, so alive is to follow its travails throughout its century of life.

This book is the biography of general relativity. Einstein's idea of how space and time are put together has taken on a life of its own, and throughout the twentieth century it was a source of delight and frustration among some of the world's most brilliant minds. General relativity is a theory that has constantly thrown up surprises, outlandish insights into the natural world that even Einstein found difficult to accept. As the theory has been passed from mind to mind, new and unexpected discoveries have come up in the strangest situations. Black holes were first conceived on the battlefields of the First World War and came of age in the hands of the pioneers of *both* the American and Soviet atom bombs. The expansion of the universe was first proposed by a Belgian priest and a Russian mathematician and meteorologist. New and strange astrophysical objects that have played a crucial role in establishing general relativity were discovered by chance. Jocelyn Bell discovered neutron stars in the Cambridge Fens using chicken wire strung out over a rickety structure of wood and nails.

The general theory of relativity was also at the heart of some of the major intellectual battles of the twentieth century. It was the target of persecution in Hitler's Germany, hounded in Stalin's Russia, and disdained in 1950s America. It has pitted some of the biggest names in physics and astronomy against each other in a battle for the ultimate theory of the universe. They slugged it out over whether the universe started with a bang or has always been eternal and what the fundamental structure of space and time really is. The theory also brought distant communities together; in the midst of the Cold War, Soviet, British, and American scientists joined to solve the problem of the origin of black holes.

The story of general relativity is not all about the past. Over the past ten years, it has become apparent that if general relativity is correct, most of the universe is dark. It is full of stuff that not only doesn't emit light but doesn't even reflect or absorb it. The observational evidence is overwhelming. Almost a third of the universe seems to be made up of dark matter, heavy, invisible stuff that swarms around galaxies like a cloud of angry bees. The other two-thirds is in the form of an ethereal substance, dark energy, that pushes space apart. Only 4 percent of the universe is made of the stuff that we are familiar with: atoms. We are insignificant. That is, *if* Einstein's theory is correct. It may just be possible that we are reaching the limits of general relativity and that Einstein's theory is beginning to crack.

Einstein's theory is also essential to the new fundamental theory of nature that has theoretical physicists at each other's throats. String theory, which attempts to go even further than Newton and Einstein and unify *everything* in nature, relies on complicated spacetimes with strange geometrical properties in higher dimensions. Far more esoteric than Einstein's theory ever was, it is hailed by many as the final theory and railed against by others as romantic fiction, not even science. Like a breakaway cult, string theory wouldn't exist if not for the general theory of relativity, yet it is looked at with skepticism by many practicing relativists.

Dark matter, dark energy, black holes, and string theory are all progeny of Einstein's theory, and they dominate physics and astronomy. While giving talks at various universities, attending workshops, and participating in meetings of the European Space Agency, responsible for some of the world's most important scientific satellites, I have come to realize that we are in the midst of a momentous transformation in modern physics. We have talented young scientists looking at general relativity with an expertise that is built on a century of geniuses. They are mining Einstein's theory with unparalleled computational power, exploring alternative theories of gravity that might dethrone Einstein's, and looking for exotic objects in the cosmos that could confirm or refute the fundamental tenets of general relativity. The wider community of scientists is simultaneously being galvanized to build colossal machines to look farther and more clearly into space than we ever have done before, satellites that will set out to search for the outlandish predictions with which general relativity seems to have burdened us.

The story of general relativity is magnificent and overarching and needs to be told. For, well into the twenty-first century, we are facing up to many of its great discoveries and unanswered questions. Something important really is going to happen in the next few years, and we need to understand where it all comes from. My suspicion is that if the twentieth century was the century of quantum physics, the twenty-first will give full play to Einstein's general theory of relativity.

Chapter 1

====

If a Person Falls Freely

D URING THE autumn of 1907, Albert Einstein worked under pressure. He had been invited to deliver the definitive review of his theory of relativity to the *Yearbook of Electronics and Radioactivity*. It was a tall order, to summarize such an important piece of work at such short notice, especially since he could do so only in his spare time. From 8:00 a.m. to 6:00 p.m. Monday through Saturday, Einstein could be found working at the Bern Federal Office for Intellectual Property in the newly built Postal and Telegraph Building, where he would meticulously pore over plans for newfangled electrical contraptions and figure out if there was any merit in them. Einstein's boss had advised him, "When you pick up an application, think that everything the inventor says is wrong," and he took his advice to heart. For much of the day, the notes and calculations for his own theories and discoveries had to be relegated to the second drawer of his desk, which he referred to as his "theoretical physics department."

Einstein's review would recap his triumphant marriage of the old mechanics of Galileo Galilei and Isaac Newton with the new electricity and magnetism of Michael Faraday and James Clerk Maxwell. It would explain much of the weirdness that Einstein had uncovered a few years before, such as how clocks would run more slowly when moving, or how objects would shrink if they were speeding ahead. It would explain his strange and magical formula that showed how mass and energy were interchangeable, and that nothing could move faster than the speed of light. His review of his principle of relativity would describe how almost all of physics should be governed by a new common set of rules.

In 1905, over a period of just a few months, Einstein had written a string of papers that were already transforming physics. In that inspired burst he had pointed out that light behaves like bundles of energy, much like particles of matter. He had also shown that the jittery, chaotic paths of pollen and dust careening through a dish of water could arise from the turmoil of water molecules, vibrating and bouncing off one another. And he had tackled a problem that had been plaguing physicists for almost half a century: how the laws of physics seem to behave differently depending on how you look at them. He had brought them together with his principle of relativity.

All these discoveries were a staggering achievement, and Einstein had made them all while working as a lowly patent expert at the Swiss patent office in Bern, sifting through the scientific and technological developments of the day. In 1907, he was still there, having yet to move into the august academic world that seemed to elude him. In fact, for someone who had just rewritten some of the fundamental rules of physics, Einstein was thoroughly undistinguished. Throughout his unimpressive academic studies at the Polytechnic Institute in Zurich, Einstein skipped classes that didn't interest him and antagonized the very people who could nurture his genius. One of his professors told him, "You are a very clever boy . . . But you have one great fault: you'll never let yourself be told anything." When Einstein's supervisor prevented him from working on a topic of his own choice, Einstein handed in a lackluster final essay, lowering his grade to a point where he was unable to secure a post as an assistant at any of the universities to which he had applied.

From his graduation in 1900 until he finally landed his job in the patent office in 1902, Einstein's career was a sequence of failures. To compound his frustration, the doctoral thesis he submitted to the University of Zurich in 1901 was rejected a year later. In his submission, Einstein had set about to demolish some of the ideas put forward by Ludwig Boltzmann, one of the great theoretical physicists of the end of the nineteenth century. Einstein's iconoclasm had not gone over well. It wasn't until 1905, when he submitted one of his magical papers, "A New Determination of Molecular Dimensions," that he finally obtained his doctorate. The degree, a newly diplomatic Einstein discovered, "considerably facilitates relations with people."

While Einstein struggled, his friend Marcel Grossmann was on the fast track to becoming an august professor. Well organized, studious, and beloved by his teachers, it was Grossmann who had saved Einstein from going off the rails by keeping detailed, immaculate notebooks of the lecture courses. Grossmann became close friends with Einstein and Einstein's future wife, Mileva Marić, while they studied together in Zurich, and all three graduated in the same year. Unlike Einstein's, Grossmann's career had progressed smoothly from then on. He had been appointed as an assistant in Zurich and in 1902 had obtained his doctorate. After a short stint teaching in high schools, Grossmann had become a professor of descriptive geometry at the Eidgenössische Technische Hochschule, known as the ETH, in Zurich. Einstein had failed to even get an appointment as a schoolteacher. It was only through the recommendation of Grossmann's father to an acquaintance, the head of the patent office in Bern, that Einstein had finally secured a job as a patent expert.

Einstein's job in the patent office was a blessing. After years of financial instability and depending on his father for an income, he was finally able to marry Mileva and begin to raise a family in Bern. The relative monotony of the patent office, with its clearly defined tasks and lack of distractions, seemed to be an ideal setting for Einstein to think things through. His assigned work took only a few hours to complete each day, leaving him time to focus on his puzzles. Sitting at his small wooden desk with only a few books and the papers from his "theoretical physics department," he would perform experiments in his head. In these thought experiments (*gedankenexperimenten* as he called them in German) he would imagine situations and constructions in which he could explore physical laws to find out what they might do to the real world. In the absence of a real lab, he would play out carefully crafted games in his head, enacting events that he would scrutinize in detail. With the results of these experiments, Einstein knew just enough mathematics to be able to put his ideas to paper, creating exquisitely crafted jewels that would ultimately change the direction of physics.

His employers at the patent office were pleased with Einstein's work and promoted him to Expert II Class, yet they remained oblivious to his growing reputation. Einstein was still working on a daily quota of patents in 1907 when the German physicist Johannes Stark commissioned Einstein to write his review "On the Relativity Principle and the Conclusions Drawn From It." He was given two months to write it, and in those two months Einstein realized that his principle of relativity was incomplete. It would need a thorough overhaul if it was to be *truly* general.

The article in the *Yearbook* was to be a summary of Einstein's original principle of relativity. This principle states that the laws of physics should look the same in any inertial frame of reference. The basic idea behind the principle was not new and had been around for centuries.

The laws of physics and mechanics are rules for how things move, speed up, or slow down when subjected to forces. In the seventeenth century, the English physicist and mathematician Isaac Newton laid out a set of laws for how objects respond to mechanical forces. His laws of motion consistently explain what happens when two billiard balls collide, or when a bullet is fired out of a gun, or when a ball is thrown up in the air.

An inertial frame of reference is one that moves at a constant velocity. If you're reading this in a stationary spot, like a cozy chair in your den or a table in a café, you're in an inertial frame. Another classic example is a smoothly moving fast train with the windows closed. If you're sitting inside it, once the train gets up to speed there's no way to know you're moving. In principle, it should be impossible to tell the difference between two inertial frames even if one is moving at a high speed and the other is at rest. If you do an experiment in one inertial frame measuring the forces acting on an object, you should get the same result as in any other inertial frame. The laws of physics are identical, regardless of the frame.

The nineteenth century brought a completely new set of laws that wove together two fundamental forces: electricity and magnetism. At first glance, electricity and magnetism appear to be two separate phenomena. We see electricity in the lights in our home or lightning in the sky, and magnetism in the magnets stuck to our fridge or the way the North Pole draws a compass. The Scottish physicist James Clerk Maxwell showed that these two forces could be seen as different manifestations of one underlying force, electromagnetism, and that how they are perceived depends on how an observer is moving. A person sitting next to a bar magnet would experience magnetism but no electricity. But a person whizzing by would experience not only the magnetism but also a modicum of electricity. Maxwell unified the two forces into one that remains equivalent regardless of an observer's position or speed.

If you try to combine Newton's laws of motion with Maxwell's laws for electromagnetism, troubles arise. If the world indeed obeys both of these sets of laws, it is possible, in principle, to construct an instrument out of magnets, wires, and pulleys that will not sense any force in one inertial frame but can register a force in another inertial frame, violating the rule that inertial frames should be indistinguishable from one another. Newton's laws and Maxwell's laws thus appear inconsistent with each other. Einstein wanted to fix these "asymmetries" in the laws of physics.

In the years leading up to his 1905 papers, Einstein devised his concise principle of relativity through a series of thought experiments aimed at solving this problem. His mental tinkering culminated in two postulates. The first was simply a restatement of the principle: The laws of physics must look the same in any inertial frame. The second postulate was more radical: In *any* inertial frame, the speed of light always has the same value and is 299,792 kilometers per second. These postulates could be used to adjust Newton's laws of motion and mechanics so that when they were combined with Maxwell's laws of electromagnetism, inertial frames remained completely indistinguishable. Einstein's new principle of relativity also led to some startling results.

The latter postulate required some adjustments to Newton's laws. In the classic Newtonian universe, speed is additive. Light emitted from the front of a speeding train moves faster than light coming from a stationary source. In Einstein's universe, this is no longer the case. Instead, there is a cosmic speed limit set at 299,792 kilometers per second. Even the most powerful rocket would be unable to break that speed barrier. But then odd things happen. So, for example, someone traveling on a train moving at close to the speed of light will age more slowly when observed by someone sitting at a station platform, watching the train go by. And the train itself will look shorter when it is moving than when it is sitting still. Time dilates and space contracts. These strange phenomena are signs that something much deeper is going on: in the world of relativity, time and space are intertwined and interchangeable.

With his principle of relativity, Einstein seemed to have simplified physics, albeit with strange consequences. But in the autumn of 1907, as Einstein set out to write the review, he had to admit that while his theory seemed to work well, it wasn't yet complete. Newton's theory of gravity didn't fit into his picture of relativity.

Before Albert Einstein came along, Isaac Newton was like a god in the world of physics. Newton's work was held up as the most stunning success of modern thought. In the late seventeenth century, he had unified the force of gravity acting on the very small and the very large alike in one simple equation. It could explain the cosmos as well as everyday life.

Newton's law of universal attraction, or the "inverse square law," is as simple as they come. It says that the gravitational pull between two objects is directly proportional to the mass of each object and inversely proportional to the square of their distance. So if you double the mass of one of the objects, the gravitational pull also doubles. And if you double the distance between the two objects, the pull *decreases* by a factor of four. Over two centuries, Newton's law kept on giving, explaining any number of physical phenomena. It proved itself most spectacularly not only in explaining the orbits of the known planets but also in predicting the existence of new ones.

Beginning in the late eighteenth century, there was evidence that the planet Uranus's orbit had a mysterious wobble. As astronomers amassed observations of Uranus's orbit, they could slowly map out its path in space with ever more precision. Predicting Uranus's orbit was not a straightforward exercise. It involved taking Newton's law of gravity and working out how the other planets influenced Uranus's motion, nudging it here and there, making its orbit ever so slightly more complicated. Astronomers and mathematicians would publish the orbits in the form of tables that would, for different days and years, predict where Uranus or any other planet should be in the sky. And when they compared their predictions with subsequent observations of Uranus's actual position, there was always a discrepancy they couldn't explain.

The French astronomer and mathematician Urbain Le Verrier was particularly skilled at working out the celestial orbits and producing orbits for various planets in the solar system. When he focused his attention on Uranus, he assumed from the start that Newton's theory *was* perfect, given how well it worked for the other planets. If Newton's theory was correct, he surmised, the only other possibility was that there had to be something out there that hadn't been accounted for. And so Le Verrier took the bold step of predicting the existence of a new, fictitious planet and producing its very own astronomical table. To his delight, a German astronomer in Berlin, Gottfried Galle, pointed his telescope in the direction that Le Verrier's table indicated and found a big, undiscovered planet shimmering in his field of view. As Galle put it in a letter to Le Verrier, "Monsieur, the planet of which you indicated the position really exists."

Le Verrier had taken Newton's theory a step further than anyone before and was rewarded for his audacity. For decades, Neptune was known as "Le Verrier's planet." Marcel Proust used Le Verrier's discovery as an analogy for ferreting out corruption in his *Remembrance of Things Past,* and Charles Dickens referred to it when describing hard-boiled detective work in his short piece "The Detective Police." It was a beautiful example of using the fundamental rules of scientific deduction. Le Verrier, basking in the glory of his discovery, then turned his attention to Mercury. It too seemed to have a strange, unexpected orbit.

In Newtonian gravity, an isolated planet orbiting the sun follows a simple, closed orbit with the shape of a squashed circle, known as an ellipse. A planet will go around and around, endlessly following the same path, periodically getting closer to and then more distant from the sun. The point in its orbit at which the planet is closest to the sun — called its perihelion — remains constant over time. Some planets, like the Earth, have almost circular orbits — the ellipse is barely squashed — while others, like Mercury, follow much more elliptical paths.

Even accounting for all the other planets' effects on Mercury's orbit, Le Verrier found that Mercury's actual orbit was at odds with the predictions of Newtonian gravity; the planet's perihelion shifted by approximately 40 arcseconds per century. (An arcsecond is a unit of angular measurement; the entire dome of the sky is made up of about 1.3 million arcseconds, or 360 degrees.) This anomaly, known as the precession of the perihelion of Mercury, could not be explained by Le Verrier's deployment of Newton's rules. Something else was going on.

Once again, Le Verrier assumed that Newton had to be right, and so, in 1859, he conjectured that a new planet, Vulcan, about the same size as Mercury had to exist very close to the sun. It was a bold, outlandish conjecture. As he put it, "How could a planet, extremely bright and always near the Sun, fail to have been recognized during a total eclipse?"

Le Verrier's conjecture set off a race to discover the new planet Vulcan. Over the following decades, there were occasional reported sightings of an object nearer the sun, but none of them stood up to scrutiny. Although the search for Vulcan didn't end with Le Verrier's death, the precession of the perihelion of Mercury remained firmly entrenched in astronomical lore. Something other than an invisible planet would have to explain the 40-arcsecond anomaly.

When Einstein sat down to worry about gravity in 1907, he had to reconcile Newton's theory with his principle of relativity. In the back of his mind, he knew that he also had to explain Mercury's anomalous orbit. It was a tall order.

Gravity as explained by Newton violates both of the postulates in Einstein's beautiful and concise principle of relativity. For a start, in Newton's theory, the effect of gravity is instantaneous. If two objects are suddenly situated near each other, the force of gravity between them would be in effect immediately — it would require no time to travel from one object to the other. But how could this be if, according to Einstein's new principle of relativity, nothing, no signal, no effect, can move faster than the speed of light? Just as crucial and as vexing was the fact that, while Einstein's principle of relativity harmonized mechanics and electromagnetism, it left out Newton's law of gravity. Newtonian gravity looked different in different inertial frames.

Einstein's first step on his long trek to fix gravity and generalize his theory of relativity came one day as he sat at his chair at the patent office in Bern, lost in his world of thought. Years later he recalled the idea that came to him and led him toward his theory for gravity: "If a person falls freely he will not feel his own weight."

Imagine yourself as Alice in the rabbit hole, falling freely with nothing to stop you. As you fall under the pull of gravity, the speed at which you fall increases at a constant rate. The acceleration will exactly match the gravitational pull, and as a result your fall will feel effortless — you won't feel any force to pull or push against — although it will be undoubtedly terrifying as you hurtle through space. Now imagine a bunch of stuff falling with you: a book, a cup of tea, a similarly panicked white rabbit. All the other objects will accelerate at the same rate to compensate for the pull of gravity, and as a result they will hover around you as you all fall together. If you try to set up an experiment with these objects to measure how they move relative to you so as to determine the gravitational force, you will fail. You will feel weightless and the objects will look weightless. All of this seems to indicate that there is an intimate relationship between accelerated motion and the pull of gravity — in this case one is exactly compensating for the other.

Maybe falling freely is a step too far. There is too much going on around you: the air is rushing by, and the fear that you'll eventually hit the bottom makes clear thinking a challenge. Let's try something slightly simpler, and a little more sedate. Imagine that you have just entered an elevator on the ground floor of a tall building. The elevator starts to go up, and in those first few seconds, as it accelerates, you feel just a little bit heavier. Conversely, suppose you are now at the top of the building and the elevator starts to go down. During those initial moments when the elevator picks up speed, you feel lighter. Of course, once the elevator reaches its maximum speed, you don't feel any heavier or lighter. But during those moments in which the elevator accelerates or decelerates, your sense of your own weight, and hence of gravity, is skewed. In other words, what you sense of gravity is completely dependent on whether you are speeding up or slowing down.

On that day in 1907 when Einstein conjured up his falling man, he realized that there must be some deep connection between gravity and acceleration that would be the key to bringing gravity into his theory of relativity. If he could change his principle of relativity so that the laws of physics remained the same not only in frames moving at constant speed but also in frames that were speeding up or slowing down, he just might be able to bring gravity into the mix with electromagnetism and mechanics. He wasn't sure how, but this brilliant insight was the initial step toward making relativity more general.

Under pressure from his German editor, Einstein wrote up his review, "On the Relativity Principle and the Conclusions Drawn From It." He included a section on what would happen if he generalized his principle to include gravity. He summarily noted a few consequences: The presence of gravity would alter the speed of light and cause clocks to run more slowly. The effects of his generalized principle of relativity might even explain the minute drift in Mercury's orbit. These effects, tossed in at the end of the paper, could eventually be used to test his idea, but they would need to be worked out in more detail and with more care at a later time. They would have to wait. For a few years, Einstein wouldn't work on his theory at all.

By the end of 1907, Einstein's brilliant obscurity was coming to an end. Slowly but surely, his 1905 papers had begun to make an impact.

He started receiving a trickle of letters from distinguished physicists asking for his offprints and discussing his ideas. Einstein was excited by the developments, telling a friend, "My papers are meeting with much acknowledgement and are giving rise to further investigation." One of his admirers quipped, "I must confess to you that I was amazed to read that you have to sit in an office for eight hours of the day. But history is full of bad jokes!" It wasn't that he had a bad life. His job in Bern had allowed him to begin a family with Mileva. In 1904 they had a son they named Hans Albert. Einstein's regular hours at the patent office allowed him to spend time at home building toys for his young son, but he was ready to enter the world of academia.

In 1908, Einstein was finally made a private lecturer at the University of Bern, a position that allowed him to give lectures to paying students. He found teaching incredibly burdensome and earned a terrible reputation as a lecturer. Still, in 1909 he was lured over to the University of Zurich as an associate professor. Einstein remained in Zurich for just over a year. In 1911, he was offered a professorship at the German University in Prague. This time he would have no teaching obligations. Without the bustle of his academic teaching duties, he returned to a state of mind much like that enabled by the ordered and isolated environment of the patent office. He could think about generalizing relativity once again.

Chapter 2

The Most Valuable Discovery

ALBERT EINSTEIN once confided to his friend and colleague the physicist Otto Stern, "You know, once you start calculating you shit yourself up before you know it." It is not that he didn't know his fair share of mathematics. Indeed, he had excelled at math in school and knew enough to put across his ideas. His papers were a perfect balance of physical reasoning and just enough mathematics to lay his ideas on a firm setting. But his 1907 predictions from his generalized theory had been done on a mathematical shoestring—one of his Zurich professors described the presentation of his work as "mathematically cumbersome." Einstein disdained mathematics, which he called a "superfluous erudition," sniping, "Since the mathematicians pounced on the relativity theory I no longer understand it myself." But in 1911, when he looked at the ideas he had written up in his review, he realized that math could help him push them a bit further.

Einstein looked at his principle of relativity and thought about light, once again. Imagine yourself riding in a spaceship far from any planets and stars. Now imagine that a ray of light from a distant star enters through a small window directly to your right, cuts across the inside of your ship, and exits through a window to your left. If your spaceship is standing still, and the light hits the window straight on, it will exit through the window directly to your left. If, however, the spaceship is moving at a very fast but constant velocity when the light ray enters, by the time the light hits the far side of the spaceship, the ship will have moved forward and the ray will exit through a window farther back on the ship. From your point of view, the light ray enters at an angle and cuts across in a straight line. If the ship is *accelerating,* things will look quite different: the light ray will *curve* through the ship and exit farther back.

Here is where Einstein's insight about the nature of gravity comes into play. Sitting in an accelerating spaceship should feel no different from sitting in a spaceship at rest, feeling the pull of gravity. As Einstein had realized, at its simplest level, acceleration is indistinguishable from gravity. Someone sitting in the spaceship as it rested on the surface of a planet would see exactly the same thing as the passenger in the accelerating craft: a light ray being bent due to gravity. In other words, Einstein realized that gravity deflects light like a lens.

The gravitational pull would have to be quite strong to actually detect any such deflection — a planet might not be enough. Einstein proposed a simple observational test using a much more massive object: to measure the deflection of distant starlight as it grazed past the edges of the sun. The angular positions of distant stars would change by a tiny amount when the sun passed in front of them, about one four-thousandth of a degree, an almost imperceptible amount that was already possible to measure with telescopes at the time. Such an experiment would have to be done during a total eclipse of the sun so that the intense brightness of its rays wouldn't overwhelm any attempts at picking out the stars from the sky.

Although Einstein had found a way to actually test the validity of his new ideas, he still couldn't make any headway in actually completing his new theory. He was still winging it on that insight he had at the patent office, the freely falling man. And although he had no teaching responsibilities and all the time in the world to conjure up his thought experiments and think more deeply about his new theory, he wasn't happy. His family had grown, with another son, Eduard, born just before he arrived in Prague, but his wife was miserable and lonely, far from the world she had grown used to in Bern and then Zurich. So Einstein jumped at the opportunity, in 1912, to move back to Zurich as a full professor at the ETH.

During his sojourn in Prague, Einstein had begun to realize that he needed a different type of language for exploring his ideas. While he was reluctant to resort to abstruse mathematics that might obscure the beautiful physical ideas he was trying to piece together, a few weeks after arriving in Zurich, he approached one of his oldest friends, the mathematician Marcel Grossmann, and pleaded, "You've got to help me or I'll go crazy." Grossmann was skeptical about the slapdash way physicists went about solving problems, but he endeavored to help his friend.

Einstein was looking at how things moved if they were accelerating or being pulled by gravity. Their paths were curves in space, not the simple, straight geometric lines that one found if one looked at inertial frames. The shape and nature of this motion were more complicated and would require Einstein to go beyond simple geometry. Grossmann gave Einstein a textbook on non-Euclidean, or Riemannian, geometry.

Almost a hundred years before Einstein started grappling with his principle of relativity, in the 1820s, the German mathematician Carl Friedrich Gauss had taken the daring step of breaking free from the geometry of Euclid. Euclid had laid down the rules for lines and shapes on flat space. Euclidean geometry is what we are still taught at school today; it is what tells us that parallel lines never intersect, and that if two straight lines intersect they do so only once. We learn that the sum of the angles in a triangle is 180 degrees, and that squares are built up of four right angles. There is a whole host of rules that we learn and apply. We draw them on flat sheets of paper and chalkboards, and they serve us well.

But what if we were asked to work on curved sheets of paper? What if we tried to draw our geometrical objects on the surface of a smooth basketball? Then our simple rules break down. For example, if we draw two lines that start off by intersecting the equator at right angles, they should be parallel. And indeed they are, but if we follow them, they end up intersecting each other at one of the poles. Hence on a sphere, parallel lines do intersect. We can go further and let these parallel lines start sufficiently far apart at the equator that they intersect at a right angle at the pole. In doing so we have constructed a triangle in which the angles add up to 270 degrees instead of 180 degrees. Again, our usual rules about triangles don't apply.

In fact, every uniquely contoured surface — a sphere, a doughnut, a crumpled piece of paper — has its own geometry with its own rules. Gauss tackled the rules for the geometry of *any* general surface you might come up with. His was a democratic view: all surfaces should be considered equal, and there should be a general set of rules for how to deal with them. Gauss's geometry was powerful and hard. It was further developed in the 1850s by another German mathematician, Bernhard Riemann, into a sophisticated and difficult branch of mathematics, so difficult that even Grossmann, who had shepherded Einstein in that direction, felt that Riemann might have gone too far for his work to be of any use to a physicist. Riemann's geometry was a mess, with many functions flying around, wrapped up in hideously nonlinear construction, but it was powerful. If Einstein could come to grips with it, he might conquer his theory.

The new geometry was fiendishly difficult, but, facing an impasse in generalizing his theory of relativity, Einstein set to work trying to master it. It was a monumental challenge, like learning Sanskrit from scratch and then writing a novel in it.

By early 1913, Einstein had embraced the new geometry and collaborated with Grossmann on two articles describing his sketch, or *Entwurf* in German, of a theory. He told one colleague, "The gravitation affair has been clarified to my full satisfaction." Couched in the new mathematics, with Grossmann writing a section explaining the new geometry to the potentially uncomprehending community of physicists, the theory incorporated the predictions that Einstein had proposed in his first forays. Einstein had succeeded in making all the laws of physics look the same in any reference frame, not just an inertial one that wasn't accelerating. He could write electromagnetism and Newton's laws of motion just as he had done in his first, more restricted theory of relativity. In fact, he had succeeded in doing so for almost all of the laws of physics *except* for gravity. The new law of gravity that Einstein and Grossmann proposed was *still* the odd one out, refusing now to yield to a general principle of relativity. Even with the new mathematics brought in to bolster his physical intuition, gravity didn't fit. All the same, Einstein was convinced he had made a major step in the right direction and just needed to tie up some loose ends before his theory would be complete. He was wrong. Einstein's final journey to his theory of spacetime would be less of a dash and more of a stumble.

In 1914 Einstein finally settled down. He was invited to Berlin to head the newly created Kaiser Wilhelm Institute of Physics, where he was to be handsomely paid and made a fellow of the august Prussian Academy of Sciences. It was the pinnacle of European academia, where he would be surrounded by brilliant colleagues such as Max Planck and Walther Nernst, and required no teaching. It was the perfect job, but it came with a personal blow. Einstein's family had had enough of all his wandering throughout Europe, and this time they didn't follow him to his new post. His wife, Mileva, remained behind in Zurich with his sons. They would remain apart for five years and divorce in 1919, and Einstein would start a new life and relationship with his younger cousin Elsa Lowenthal, whom he would marry in 1919 and with whom he would remain until her death in 1936.

Einstein arrived in Berlin at the beginning of the Great War and found himself caught up "in the madhouse," as he put it, of German nationalism. It affected almost everyone. All around him, colleagues were going to the front or developing new weapons for the battlefield such as the dreaded mustard gas. In September 1914 a nationalist manifesto, "An Appeal to the Cultured World," came out, supporting the German government. Signed by ninety-three German scientists, authors, artists, and men of culture, it set out to counterattack the misinformation propagated about Germany throughout the world. Or so they thought. The manifesto claimed that Germans were not responsible for the war that had just broken out. It conveniently glossed over the fact that Germany had just invaded Belgium and devastated the city of Louvain, simply stating that "the life or property of [not] even a single Belgian citizen was touched by our soldiers." It was defiant and divisive, and much of it wasn't true.

Einstein was shocked by what was going on around him. As a pacifist and internationalist, he entered the fray with a countermanifesto, "An Appeal to Europeans." In it, Einstein and a handful of colleagues distanced themselves from the "Manifesto of the Ninety-three," firmly chastising their colleagues and entreating the "educated men of all states" to fight against the destructive war around them. The "Appeal to Europeans" was, on the whole, ignored. To the outside world Einstein was just another of the German scientists who supported the document of the ninety-three, and hence he was the enemy. At least that was the view from England.

The Englishman Arthur Eddington was known for cycling long distances. He had devised a number, E, that encapsulated his cycling stamina. E was, put simply, the largest number of days in his life that he had cycled more than E miles. I doubt I have an E number greater than 5 or 6. I haven't biked six miles in a day more than six times in my life — a pathetic number, I know. When Eddington died, he had an E number of 87, which means he had taken eighty-seven individual bike rides that were longer than eighty-seven miles. His unique stamina and perseverance served him well and would push him to achieve quite spectacular results in all walks of life.

Whereas Einstein had struggled to begin his scientific career, Eddington had been fast-tracked into the heart of English academia. Eddington could be arrogant, dismissive, and disconcertingly stubborn when promoting his own ideas, but he was also a tenacious scientist who was rarely put off by fiendishly difficult astronomical observations or new esoteric mathematics. He had been brought up in a devout Quaker family and from early on had excelled at school. At sixteen he went to Manchester to study mathematics and physics and ended up in Cambridge, where he was the top-scoring student of his year, known as the "Senior Wrangler." On finishing his MA, he was almost immediately made an assistant to the Astronomer Royal and fellow of Trinity College, Cambridge.

Cambridge was high octane, and Eddington was surrounded by brilliant scholars. There was J. J. Thomson, who had discovered the electron, and A. N. Whitehead and Bertrand Russell, who together had written the *Principia Mathematica,* a true bible for logicians. Over time he would be joined by Ernest Rutherford, Ralph Fowler, Paul Dirac, and a veritable who's who of twentieth-century physics. Eddington fit right in. After spending a few years at the Greenwich Observatory in London, he returned to Cambridge. At only thirty-one years of age, he was appointed to the prestigious position of Plumian Professor of Astronomy and Experimental Philosophy at the University of Cambridge. He was also appointed director of the Cambridge Observatory on the outskirts of town, and he settled there with his sister and his mother to become the leader of British astronomy. Eddington would remain there for the rest of his life, taking part in college life with its formal dinners and staid debates, regularly visiting the Royal Astronomical Society to present his results, and every now and then traveling to some far corner of the world to make measurements and observe the skies.

It was on one such trip that Eddington first came across Einstein's new ideas on gravity. Einstein's proposed bending of light had caught the fancy of a few astronomers who had taken it upon themselves to try to measure it. They would set off across the globe, to America, Russia, and Brazil, trying to capture an eclipse at just the right moment and with the sun in the right position so that they could measure the slight deflections of distant stars. While observing an eclipse in Brazil, Eddington met one such astronomer, the American Charles Perrine, and was intrigued by what he was doing. So when he returned to Cambridge, Eddington decided to look into Einstein's new ideas.

When the Great War broke out, Eddington was one of the lone voices opposing the wave of rabid nationalism that was subsuming not only his country but his colleagues. It drove him to despair. In a series of angry pieces in *The Observatory,* the mouthpiece of British astronomers, the case against working with German scientists was made forcefully by a slew of senior astronomers. The Savilian Professor of Astronomy at Oxford, Herbert Turner, put it succinctly: "We can readmit Germany to international society and lower our standards of international law to her level or we can exclude her and raise it. There is no third way." Such was the animosity against anything German that the president of the Royal Astronomical Society, who had a German background, was asked to resign. British scientists' relations with their German colleagues were frozen for the duration of the war.

Eddington thought and behaved differently. As a Quaker he was passionately opposed to war. During the mounting anger against the German intelligentsia, he found himself speaking out in dissent. "Think, not of a symbolic German, but of your former friend Prof. X, for instance," he appealed to his colleagues. "Call him Hun, pirate, baby-killer, and try to work up a little fury. The attempt breaks down ludicrously." Eddington not only spoke out for the Germans; he refused to be sent into battle and to fight. As he witnessed some of his friends and colleagues being shipped off to the front to be killed in action, Eddington campaigned against the war. Given an exemption out of "national importance"— he was more important to the nation as an astronomer than as a foot soldier — he made few friends.

Alone in Berlin, surrounded by the mayhem of war, Einstein worked on perfecting his final theory. It looked correct, but he needed more math to make it right. So he set off to the University of Göttingen, then the mecca of modern mathematics, to visit the mathematician David Hilbert. Hilbert was a colossus and ruled the world of mathematicians. He had transformed the field, attempting to lay down an unshakable formal foundation from which all of mathematics could be constructed. There would be no more looseness in mathematics. Everything would have to be deduced from a basic set of principles using well-established formal rules. Mathematical truths were *really* truths only if proved according to these rules. This had become known as the "Hilbert Program."

Hilbert had surrounded himself with some of the most important mathematicians in the world. One of his colleagues had been Hermann Minkowski, who had shown Einstein how his special theory of relativity could be written in a far more elegant, mathematical language — the "superfluous erudition" that Einstein had disparaged a few years before. Hilbert's students and assistants — such as Hermann

Weyl, John von Neumann, and Ernst Zermelo — would be leading fig-
ures in twentieth-century mathematics. Along with his group at Göttingen, Hilbert had grand plans: to construct a complete theory of the
natural world based on first principles, just as in mathematics. He saw
Einstein's work as an integral part of his project.

During Einstein's short visit to Göttingen in June of 1915, Einstein
lectured and Hilbert took notes. They discussed and argued back and
forth about the details. Einstein was strong on the physics and Hilbert
on the mathematics. But they didn't make any progress. Einstein, still
wary of mathematics and still shaky in his understanding of Riemannian geometry, found it difficult to completely understand Hilbert's detailed, technical points.

Shortly after Einstein ended what seemed like a fruitless visit, he
began to doubt his new theory of relativity. He already knew that it
wasn't truly general — when he and Grossmann had finished their papers in 1913, it was clear to him that the law of gravity *still* didn't fit. And
some of his predictions were off. For example, his theory predicted a
drift for Mercury, very much as Le Verrier had observed almost fifty
years before, but it wasn't *exactly* right. It was still off by a factor of two.
Einstein had to look at his equations again.

Over a period of just three weeks, Einstein decided to ditch the new
law of gravity that he had proposed with Grossmann, which didn't
obey the general principle of relativity. He wanted a law of gravity that
would be true in any reference frame, much as he had already done
with the other laws of physics. And he wanted to use the new Riemannian geometry that he had learned from Grossmann. Every few
days he would tweak what he had done before, writing down a law,
relaxing some assumptions while imposing others. And as he did so,
he shed some of the physical prejudices that had held him back and
delved deeper and deeper into the mathematics that he had learned.
He realized that even though his physical intuition had served him
well throughout his spectacular career, he had to be careful not to let it
cloud the bigger picture coming out of the mathematics.

Finally, by the end of November, he realized he had done it. He had
finally discovered a general law for gravity that satisfied the general
principle of relativity. On the scale of the solar system it was accurately

approximated by Newtonian gravity, exactly as it should be. Moreover, it predicted Le Verrier's precession of the perihelion of Mercury bang-on. And it predicted that as light rays passed by a heavy object, they would be bent even more — in fact twice as much as he had originally predicted when he first thought of the idea in Prague.

Einstein's completed general theory of relativity offered an entirely new way of understanding physics, one that superseded the Newtonian view that had held sway for centuries. His theory provided a set of equations that came to be known as the "Einstein field equations." Although the idea behind them, relating the geometry of Gauss and Riemann with gravity, was beautiful — "elegant," as physicists would want to call it — the detailed equations could look like a mess. They were, in practice, a set of ten equations of ten functions of the geometry of space and time, all nonlinearly tangled up and intertwined such that, in general, it was impossible to solve for one function at a time. They all had to be tackled together, head-on — a truly daunting prospect. Yet they held much promise, for their solutions could be used to predict what would happen in the natural world, from the motion of a bullet or an apple falling off a tree to the movement of planets in the solar system. The secrets of the universe, it seemed, were to be found by solving Einstein's equations.

On November 25, 1915, Einstein presented his new equations to the Prussian Academy of Sciences in a short three-page paper. His new law of gravity was radically different from what anyone had ever proposed before. In essence, Einstein argued that what we perceive as gravity is nothing more than objects moving in the geometry of spacetime. Massive objects affect the geometry, curving space and time. Einstein had finally arrived at his truly general theory of relativity.

But Einstein was not alone. Hilbert had been mulling over Einstein's Göttingen lectures and had, without Einstein realizing, made his own attempts at coming up with new gravitational equations. Completely independently, Hilbert had come up with exactly the same gravitational law. On the twentieth of November, five days before Einstein's presentation to the academy in Berlin, Hilbert presented his own results to the Royal Society of Sciences in Göttingen. It seemed as if Hilbert had scooped Einstein.

During the weeks following the announcements, relations between Hilbert and Einstein were strained. Hilbert wrote to Einstein claiming he didn't remember the bit in one of his lectures in which Einstein had discussed his attempts at building the gravitational equations, and by Christmas, Einstein was satisfied that there hadn't been any foul play. As Einstein said, in a letter to Hilbert, to begin with "there has been between us something like a bad feeling," but he had come to terms with what had happened, so much so that "I once more think of you in unclouded friendship. . . ." They would indeed remain friends and colleagues, for Hilbert stepped back from claiming any credit for Einstein's magnum opus. In fact, until he died, Hilbert always referred to the equations that both he and Einstein had stumbled upon as "Einstein's equations."

Einstein had completed his trek. He had gradually succumbed to the power of mathematics to reach his final equations. From then on he would let himself be guided not only by his thought experiments but also by the mathematics. The sheer mathematical beauty of his final theory stunned him. He described his equations as "the most valuable discovery of my life."

Eddington had been receiving the slow trickle of offprints coming out of Prague, then Zurich, and finally Berlin from a friend, the astronomer Willem de Sitter, from Holland. He was intrigued, hooked by this completely new way of looking at gravity in a difficult language. Even though he was an astronomer, and his job was to measure and observe things and try to interpret them, he was up to the challenge of learning the new mathematics of Riemannian geometry that Einstein had used to write up his theory. And it was well worth looking into, especially since Einstein had made quite clear predictions that could be used to test his theory. In fact, an eclipse was predicted to occur on the twenty-ninth of May, 1919, an ideal opportunity for such a test, and Eddington would be the obvious person to lead a team of observers.

There was only one problem, and a massive one at that. Europe was at war, Eddington was a pacifist, and Einstein was in cahoots with the enemy. Or so Eddington's colleagues wanted him to believe. As the war reached its climax in 1918, the risk of the German army completely engulfing the British and the French grew, leading to a renewed wave of conscriptions. Eddington was called up to fight, but he had something else on his mind.

While Eddington had become an enthusiastic advocate of Einstein's new gravity, he faced the antipathy of his colleagues. In an attempt to dismiss German science as having no worth, one of his colleagues declared, "We have tried to think that exaggerated and false claims made by Germans today were due to some purely temporary disease of quite recent growth. But an instance like this makes one wonder whether the sad truth may not lie deeper." And while Eddington had the support of the Astronomer Royal, Frank Dyson, to lead the eclipse expedition, he had to escape being sent to jail for refusing to fight. The British government convened a tribunal in Cambridge to look into Eddington's stance. As the hearing proceeded, the tribunal viewed him with increasing hostility. Eddington was going to be refused exemption until Frank Dyson stepped in. Eddington was a crucial player in the eclipse expedition, Dyson said, and furthermore, "under present conditions the eclipse will be observed by very few people. Prof. Eddington is peculiarly qualified to make these observations, and I hope the Tribunal will give him permission to undertake this task." The eclipse intrigued the tribunal, and Eddington was once again given an exemption for "national importance." Einstein had saved him from the front.

From Einstein's theory there was a prediction: that the light emitted from distant stars would bend as it passed close to a massive body such as the sun. Eddington's experiment proposed to observe one such distant cluster of stars, the Hyades, at two different times of the year. He would first accurately measure the positions of the stars in the Hyades cluster on a clear night, with nothing obscuring his view and nothing in the way to bend their light rays. Then he would measure their position again, this time with the sun in front. It would have to be done during a total eclipse, when almost all the bright light of the sun would be blocked by the moon. On the twenty-ninth of May, 1919, the Hyades would lie right behind the sun and conditions would be perfect. A comparison of the two measurements — one with the sun and one without — would show if there was any deflection. And if that deflection was about four-thousandths of a degree, or 1.7 arcseconds, it would be just as Einstein claimed. Such a clear and simple goal.

It wasn't actually that simple. The few places on the Earth where one could witness the total eclipse were remote and far apart. The astronomers would need to travel quite far, in a world that had just come out of a devastating war, to set up their equipment. Eddington, along with Edward Cottingham from the Greenwich Observatory, set up shop on the island of Príncipe. A backup team of two astronomers, Andrew Crommelin and Charles Davidson, was dispatched to a village called Sobral in the interior of the Brazilian Nordeste, a poor, dusty region near the equator.

Príncipe is a small island in the Gulf of Guinea, a Portuguese colony known for its cocoa. A lush green island, hot, humid, and periodically peppered with tropical storms, it had a few large *roças,* or plantations, spread out where a few Portuguese landowners used the local inhabitants to farm the land. For decades it had supplied the cocoa beans to the Cadbury corporation. At the beginning of the twentieth century the cacao plantations were accused of using slave labor and lost their contracts, destroying Príncipe's economy. When Eddington arrived, the island was slipping into oblivion.

Eddington set up his apparatus in a remote corner at the Roça Sundy, where he was looked after by the landowner. Between daily tennis matches on the only court on the island, he waited for the day of the eclipse, praying that the recurring rainstorms and gray skies wouldn't sabotage his mission. Cottingham primed the telescope, hoping that the heat wouldn't distort the images.

On the morning of the eclipse, it rained heavily and the sky was completely impenetrable until, less than an hour before totality, it started to clear. Eddington and Cottingham caught their first glimpse of the eclipse that was under way with part of the sun already obscured. By 2:15 in the afternoon, the sky was clear and Eddington and Cottingham could take their measurements — sixteen photographic plates of the sun with the Hyades cluster lurking in the background. By the end of the eclipse, the sky was beautiful, clear of any cloud. Eddington telegraphed a message to Frank Dyson: "Through cloud. Hopeful."

The cloudy start to the experiment in Príncipe may have saved the day. In Sobral in the Brazilian Nordeste, there was a perfectly clear and hot day on which the eclipse could be followed right from the very start. Crommelin and Davidson were surrounded by the jubilant locals to witness the historic event and were able to take nineteen plates to complement the sixteen taken by Eddington and Cottingham. Exultant, they also telegraphed back: "Eclipse Splendid." At the time, they didn't realize that the good viewing conditions and hot clear weather in Brazil had sabotaged their main experiment. The heat had warped the apparatus so much that the measurements on the photographic plates were rendered useless. It was only with backup observations with a smaller telescope that the expedition to Sobral was able to contribute data to the experiment.

The astronomers were unable to return home quickly, and it was only in late July that the various photographic plates began to be analyzed. Of the sixteen plates that Eddington had recorded, only two had enough stars to measure the deflection properly. The value they got was 1.61 arcseconds with an error of 0.3 arcseconds, consistent with Einstein's prediction of 1.7 arcseconds. When the plates from Sobral were analyzed, the results were worrying. The value measured was 0.93 arcseconds, far from the relativistic prediction and very close to the Newtonian prediction, but these were the same plates that had been deformed by the heat. When the backup observations from Sobral, undertaken on the smaller telescope, were analyzed, the deflection came out at 1.98 arcseconds with a very small error of 0.12 arcseconds. Einstein's prediction, again.

On November 6, 1919, the team of explorers presented their results to a joint meeting of the Royal Society and the Royal Astronomical Society. In a series of talks led by Frank Dyson, the different measurements from the eclipse expedition were laid out in front of an audience of their distinguished peers. Once the problems that had faced the Sobral expedition were taken into account, the speakers showed that the eclipse measurements spectacularly confirmed Einstein's prediction.

J. J. Thomson, the president of the Royal Society, described the measurements as "the most important result obtained in connection with the theory of gravitation since Newton's day." He added, "If it is sustained that Einstein's reasoning holds good — and it has survived two very severe tests in connection with the perihelion of Mercury and the present eclipse — then it is the result of one of the highest achievements in human thought."

The day after the Burlington House meeting, Thomson's words appeared in the London *Times*. Next to a clutch of headlines celebrating the anniversary of the armistice and praising the "Glorious Dead" was an article with the headline "Revolution in Science. New Theory of the Universe. Newton's Ideas Overthrown," describing the results from the eclipse expeditions. News and opinions about Einstein's new theory and Eddington's expedition spread like wildfire through the English-speaking world. By the tenth of November news had reached America, where the *New York Times* published its own eye-catching headlines: "All Lights Askew in the Heavens," "Einstein's Theory Triumphs," and the more convoluted "Stars Not Where They Seemed or Were Calculated to Be but Nobody Need Worry."

Eddington's gamble had paid off. By testing and actually understanding Einstein's new general theory of relativity, he had established himself as the prophet of the new physics. From then on, Eddington would be one of the few pundits to whom everyone would defer when discussing the new relativity, and his opinions would be sought, above anyone else's, as a guide to how Einstein's theory should be interpreted or developed.

And, of course, Eddington's spectacular mission had made Einstein a superstar. His findings would transform Einstein's life and propel his general theory of relativity, at least for a while, to a level of popularity and fame rarely experienced by a scientist. He had dethroned Newton, who had reigned supreme for hundreds of years. Even though his theory was opaque and couched in a mathematical language that very few people understood, it had passed Eddington's test with flying colors. Furthermore, Einstein had stopped being the enemy. The war was over, and while a lingering animosity against the German scientists remained, Einstein was excused. It was now publicly known that he hadn't signed the Manifesto of the Ninety-three, and in fact he wasn't even a German, but a Swiss Jew. As Einstein wrote in an article in the

Times shortly after Eddington's historic announcement at the RAS, "In Germany I am called a German man of science, and in England I am represented as a Swiss Jew. If I come to be represented as a bête noire, the descriptions will be reversed, and I shall become a Swiss Jew for the Germans and a German man of science for the English."

From being an unknown patent clerk, with a tendency toward insolence, admired by a few specialists in his field, Einstein had become a cultural icon, invited to give lectures in America, Japan, and throughout Europe. And his general theory of relativity, which had first seen the light of day in a simple thought experiment in his office in Bern, was now fully formed as a new, completely different way of doing physics. Mathematics had taken a firm foothold in the physics of relativity, resulting in a set of intricate and beautiful equations that were ready to be let loose on the world. It was time for others to start figuring out what they meant.

Chapter 3

―――――――――

Correct Mathematics, Abominable Physics

E INSTEIN'S FIELD EQUATIONS were complicated, a tangle of many unknown functions, yet they could in principle be solved by anyone with the right ability and determination. In the decades that followed Einstein's discovery, an eclectic Soviet mathematician and meteorologist named Alexander Friedmann and Abbé Georges Lemaître, a brilliant, determined Belgian priest, took the equations of general relativity and constructed a radical new view of the universe, a view that Einstein himself refused to accept for a very long time. Through their work, the theory gained a life of its own, beyond Einstein's control.

When Einstein first formulated his field equations in 1915, he had wanted to solve them himself. Finding a solution to his equations that could accurately model the whole universe seemed a good place to start. In 1917 he set about doing so, making some simple assumptions. In Einstein's theory, the distribution of matter and energy told spacetime what to do. To model the universe as a whole, he needed to consider *all* the matter and energy in the universe. The simplest and most logical assumption, and the one Einstein adopted in his first attempt, was that matter and energy are spread evenly throughout the whole of space. In doing so, Einstein was just continuing a line of reasoning that had transformed astronomy in the sixteenth century. Then, Nicolaus Copernicus had made the brave proposal that the Earth wasn't the center of the cosmos and that, in fact, it orbited around the sun. This "Copernican" revolution had succeeded throughout the centuries in making our place in the cosmos ever more insignificant. By the mid-nineteenth century, it became clear that not even the sun was of great import and lay somewhere nondescript in one of the spiral arms of the Milky Way, our galaxy. When Einstein tackled his equations, he was merely extending the idea that anywhere in the universe should look more or less the same to its logical consequences: there should be no preferred place or center that stands out.

The assumption that the universe was full of stuff, evenly spread out, made the field equations much simpler, but it also led to a very strange result: Einstein's equations predicted that such a universe would start to evolve. At some point, all the evenly distributed bits of energy and matter would start moving relative to each other in an organized manner. On the largest scales, nothing would stay still. Eventually everything could even fall in on itself, pulling spacetime along with it and causing the entire universe to collapse out of existence.

In 1916, astronomers' general view of the cosmos was parochial at best. While they had a pretty good map of the Milky Way, there was little, if any, sense of what lay beyond it. No one had a clear indication of what the universe was doing as a whole. All observations seemed to show that stars were moving about a little bit, but not dramatically and definitely not in a concerted, organized manner on a large scale. To Einstein, as to most people, the sky seemed static, and there was no evidence that the universe was collapsing or expanding. Letting his physical intuition and prejudice get the better of him, Einstein proposed a fix to eradicate the evolving universe from his theory. He attached a new constant term to his field equations. This cosmological constant would stabilize the universe by exactly compensating for all the stuff in it. All the ordinary stuff, the energy and matter that Einstein had spread out evenly in the universe, tried to pull spacetime in on itself, and the cosmological constant pushed back, preventing the universe from collapsing. This push and pull kept the universe in a delicate, balanced state: fixed and static, exactly as Einstein believed it should be.

Shying away from the conclusion that the universe was evolving immensely complicated Einstein's own theory. As he himself would later admit, "The introduction of such a constant implies a considerable renunciation of the logical simplicity of the theory." By adding the constant, he told a friend he had "committed something in the theory of gravitation that threatens to get me interned in a lunatic asylum." But it did the job.

In the crescendo that led up to the discovery of relativity, Einstein would often write and discuss his work with Willem de Sitter, a Dutch astronomer at Leiden University, in Holland. Living in a neutral country during the First World War, de Sitter had been instrumental in relaying information about Einstein's theory to England, where Eddington had studied his work in detail; de Sitter was the quiet man who had played a pivotal role in the lead-up to the 1919 eclipse expedition.

A mathematician by training, de Sitter was well equipped to tackle the Einstein field equations. The moment he received a draft of Einstein's paper describing a static universe born out of the field equations mangled with the cosmological constant, de Sitter realized that Einstein's solution was not the only possibility. In fact, he pointed out, it was possible to construct a universe containing nothing but the cosmological constant. He proposed a realistic model of a universe that could contain stars, galaxies, and other matter, but in such small quantities that they would have no effect on spacetime and would be unable to balance out the cosmological constant. As a result, the geometry of de Sitter's universe would be completely determined by Einstein's fix, the cosmological constant.

Both Einstein's and de Sitter's universes were static and unevolving, exactly as Einstein's prejudices had led him to believe. Yet de Sitter's universe had a strange property that de Sitter himself noted in his papers. De Sitter had built his universe so that spacetime was static, just as Einstein had before him. The universe's geometry, such as how curved space was at each point, would remain unchanged over time. But if you now scattered a few stars and galaxies in de Sitter's universe—a reasonable thought exercise given that our own universe seems to be full of such things—they would all start to move in a concerted manner, drifting away from the universe's center. Even though the *geometry* in de Sitter's universe was completely static and stayed the same for all time, objects within his universe wouldn't stay still.

A few weeks after receiving Einstein's paper describing his static universe, de Sitter had already written up his own solution and sent it back to Einstein. While Einstein recognized that de Sitter's model was mathematically valid, he was not impressed and he hated the idea of a universe completely empty of the planets and stars that we can see in the night sky. For Einstein, all that stuff was essential and was what made us have a sense that we were moving or turning. Only relative to the firmament of stars could we say if we were accelerating, slowing down, or spinning. They gave us a reference for applying all the laws of physics. Without all that stuff, Einstein's intuition failed him. He wrote back to Paul Ehrenfest expressing his irritation at this world devoid of matter. "To admit such possibilities," he wrote, "seems senseless." Despite Einstein's grumbling, within just a few years of its creation, general relativity had spawned two static models of the universe that were very different at their core.

While Einstein was working on his general theory of relativity, Alexander Friedmann was bombing Austria. As a pilot for the Russian army, Friedmann had volunteered in 1914, serving first in an air reconnaissance unit on the northern front and later on in Lvov. For a short while, it almost seemed that the Russians would prevail against the enemy. On regular night flights over southern Austria, he would join his colleagues in bringing towns that were blockaded by the Russian army into submission. Town by town, the occupying Russians were taking control.

Friedmann was different from the other pilots. While his colleagues dropped their bombs by eye, making rough guesses of where they would land, Friedmann was more careful. He had come up with a formula that would take into account his speed, the bomb's velocity, and its weight and would predict where he had to drop it to hit the desired target. As a result, Friedmann's bombs always hit their marks. He was awarded the Cross of St. George for his bravery in combat.

Having specialized in pure and applied mathematics before 1914, Friedmann had a great talent for calculation. He often threw himself into problems that were too difficult to solve exactly in the era before computers. Friedmann was fearless and would strip his equations down to their bare essentials, simplifying the messiness wherever he could and getting rid of any extra baggage. If he still couldn't solve them, he would draw graphs and pictures that would gently approximate the right results, giving him the answers he wanted. With a voracious appetite for solving problems, Friedmann tackled everything, from weather forecasting to the behavior of cyclones and the flow of fluids to the trajectories of his bombs. He was undaunted by difficulty.

At the beginning of the twentieth century, Russia was changing. The Tsarist regime lurched from crisis to crisis, ill equipped to deal with the growing discontent among a hugely impoverished population and facing the increasing turmoil in an ever more unstable Europe. Friedmann was enthusiastic about playing a part in the social changes around him. As a high school student, he fought alongside his fellow students during the first Russian Revolution of 1905, leading some of the school protests that shook the country. As an undergraduate at Saint Petersburg University he stood out for his brilliance, and during the war he led from the front, flying, bombing, teaching aeronautics, and running an industrial plant for producing navigational instruments.

After the war, Alexander Friedmann settled as a professor in Petrograd (later to become known as Leningrad). The "relativity circus," as Einstein called it, had arrived in Russia. Intrigued by the weird and wonderful mathematics, Friedmann decided to deploy his formidable mathematical skills in attempting to solve Einstein's equations. Just as Einstein had done before him, Friedmann untangled the complicated knot of equations by assuming that the universe was simple on the largest scales, that matter was distributed evenly, and that the geometry of space could be described solely in terms of one number, its overall curvature. Einstein had argued that this number was fixed once and for all as a result of the delicate balance between his cosmic term, the cosmological constant, and the density of matter, in the form of stars and planets sprinkled through space.

Friedmann ignored Einstein's results and started from scratch. By studying how matter and the cosmological constant affected the geometry of the universe, he came up with a startling fact: that one number, the overall curvature of space, evolved with time. The ordinary stuff in the universe, the stars and galaxies sprinkled all over the place, would cause space to contract and fall in on itself. If the cosmological constant was a positive number, it would push space apart, making it expand. Einstein had balanced these two effects against each other, the pulling and the pushing, so that space stayed still. But Friedmann found that this static solution was only a particular special case. The general solution was that the universe *had to* evolve, contracting or expanding depending on whether matter or the cosmological constant played the dominant role.

In 1922, Friedmann published his seminal paper, "On the Curvature of Space," in which he showed that not only Einstein's but also de Sitter's universes were merely very special cases of a much wider range of possible behaviors for the universe. In fact, the most general solutions were for universes that either contracted or expanded in time. A certain class of models could even expand and grow and then contract again, leading to a never-ending succession of cycles. Friedmann's results also released Einstein's cosmological constant from its duty of keeping the universe static. There was nothing to pin the cosmological constant to any particular value, unlike in Einstein's original model. In the conclusions of his paper, Friedmann wrote dismissively, "The cosmological constant . . . is undetermined . . . since it is an arbitrary constant." By giving up Einstein's requirement that the universe be static, Friedmann had shown that Einstein's cosmological constant was, to all effects, irrelevant. If the universe evolved, there was no need to complicate the theory with an arbitrary fix as Einstein had done.

Here was a paper that came out of nowhere. Friedmann had not taken part in the discussions with Einstein, had not sat through the succession of lectures that Einstein had given to the Prussian Academy of Sciences. He was an outsider who had become enthused by the wave of euphoria that had followed Eddington's eclipse expedition. A mathematical physicist first and foremost, all Friedmann had done was deploy the same skills and techniques he had used for studying bombs and the weather, and he had uncovered a result that went against Einstein's gut feeling.

For Einstein, the possibility that the universe was evolving was absurd. When Einstein first read Friedmann's paper, he refused to accept that his theory would serve up such a possibility. Friedmann *must* be wrong, and Einstein set about trying to prove it. He carefully worked through Friedmann's paper and found what he took to be a fundamental mistake. Once that mistake was corrected, Friedmann's calculation delivered up a static universe just as Einstein had predicted. Einstein rapidly published a note in which he asserted that "the significance" of Friedmann's work was to prove that the universe's behavior was constant and immutable.

Friedmann was mortified by Einstein's note. He was sure he hadn't made a mistake and that Einstein himself had miscalculated. Friedmann wrote a letter to Einstein showing where Einstein had gone wrong and added at the end: "If you find the calculations presented in my letter correct, please be so kind as to inform the editors of the *Zeitschrift für Physik* about it." He sent off his letter to Berlin, hoping Einstein would act swiftly.

Einstein would never receive the letter. His fame had propelled him into an endless succession of seminars and conferences, forcing him to travel around the world, from Holland and Switzerland to Palestine and Japan, and keeping him away from Berlin where Friedmann's letter sat gathering dust. It was only by chance that Einstein ran into one of Friedmann's colleagues while passing through the Leiden Observatory and learned about Friedmann's response. And so it was that, almost six months later, Einstein published a correction to *his* correction of Friedmann's paper, rightfully acknowledging Friedmann's main result and admitting "there are time varying solutions" to the universe. The universe could indeed evolve in his general theory of relativity. But still, all Friedmann had done was show that there were solutions to Einstein's theory that led to an evolving universe. That was just mathematics, according to Einstein, not reality. His prejudice still led him to believe that the universe had to be static.

Friedmann gained notoriety for having corrected the great man himself. But even though he set some of his doctoral students to extend his ideas even further, and he himself continued publicizing Einstein's work throughout what had by then become known as the Soviet Union, he returned to his work on meteorology. Friedmann died in 1925, at the age of thirty-seven, from typhoid fever caught while he was on holiday in Crimea, and his mathematical model of an evolving universe was to lie dormant for a number of years.

Georges Lemaître came to math and religion at a young age. He was good with equations, clever at coming up with clean, new solutions to the mathematical conundrums he was set in school. Having attended a Jesuit school in Brussels, Lemaître went on to study mining engineering and was still doing so when he was called up for the war in 1914. While Einstein and Eddington were campaigning for peace, Georges Lemaître was fighting in the trenches when the Germans invaded Belgium. The Germans destroyed the city of Louvain and outraged the international community, leading to the infamous manifesto of the ninety-three German scientists that so poisoned relations between English and German science. Lemaître was an exemplary soldier, becoming a gunner and rising in the ranks to become an artillery officer. Like Alexander Friedmann, he applied his knack for solving intricate problems to ballistics. When the war ended, he was cited for bravery in the Belgian army's Orders.

Lemaître's experience of the carnage of battle, the devastating effect of chlorine gas in the trenches, and the brutality of the front affected him profoundly. Following active duty, he not only studied physics and mathematics but also entered the Maison Saint Rombaut in 1920 and by 1923 had been ordained a Jesuit priest. For the rest of his life, Lemaître would pursue his fascination for mathematics alongside his spiritual devotion, rising through the ranks of the Catholic Church to become the president of the Pontifical Academy of Sciences. He was a scientist priest who would turn his sights to solving the equations of the universe.

While at university, Lemaître had already been enticed by Einstein's general theory of relativity, giving seminars and writing short reviews on the topic at the University of Louvain. In 1923, he spent time in Cambridge, England, boarding at a house for Catholic clergymen and working with Eddington on relativity. Eddington pointed Lemaître to the foundations of relativity, giving him a front-row seat as the search for the true theory of the universe unfolded. Eddington was impressed by Lemaître, finding him "a very brilliant student, wonderfully quick and clear-sighted, and of great mathematical ability." When Lemaître moved to Cambridge, Massachusetts, in 1924, the unsolved problem of how to accurately model the universe became his main concern, one he delved into deeply as he worked on his PhD at MIT.

When Lemaître turned to cosmology in 1923, the two world models of Einstein and de Sitter were still at play. They were still the only two mathematical models to have come out of Einstein's equations, yet they remained just that: two mathematical models without any observations privileging one above the other. Alexander Friedmann's evolving universe had failed to make any impact, and Einstein's prejudice against an evolving universe held enough weight to prevent anyone from pursuing it. According to the prevailing view, the universe was still very static. But Eddington had been intrigued by de Sitter's model, in which stars and galaxies drifted away from the center of the universe. De Sitter had argued there might be a distinct observational signature of his universe. In such a universe, distant objects would look peculiar. Their light would be redshifted.

We can think of light as a collection of waves with different wavelengths corresponding to different energy states. Red light has a longer wavelength and lower energy state than blue light, at the other end of the spectrum. When we look at a star or galaxy, or any bright object, the light it emits is a mixture of these waves, some more energetic than others. What de Sitter found was that the light of any faraway object would be invariably pushed toward the red, appearing to have a longer wavelength and less energy than similar objects nearby. The farther away an object was, the redder it would be. A sure way to test de Sitter's model would be to look for this phenomenon in the real universe.

The redshift effect, in which distant galaxies seemed to be more redshifted than closer ones, hinted that there was something not completely understood about de Sitter's model. With Hermann Weyl, one of David Hilbert's disciples from Göttingen, Eddington examined de Sitter's solution more closely and found that if one sprinkled stars or galaxies all over spacetime, a very tight, linear relationship between the redshifts and distances of each star or galaxy emerged. An object that was twice as far from Earth as another would have a redshift that was correspondingly twice as large. This pattern of redshifting became known as the de Sitter effect.

When, in 1924, Lemaître took a closer look at de Sitter's universe and Eddington and Weyl's findings, he realized the equations in de Sitter's paper were written in an odd way. De Sitter had formulated his theory using a static universe with a strange property: his universe had a center, and for an observer positioned at its center, there was a horizon beyond which nothing could be seen. This was at odds with one of Einstein's basic assumptions about the universe, that all places were equal. When Lemaître reformulated de Sitter's universe so that the horizon went away and all points in space were considered equal, he found that the de Sitter universe behaved in a completely different way. Now, in Lemaître's simpler way of looking at the universe, the curvature of space evolved with time and the geometry evolved as if points in space were hurtling away from each other. It was this evolution that could explain the de Sitter effect. Just like Friedmann a couple of years before, Lemaître had stumbled upon the evolving universe. Lemaître's discovery that redshift was associated with an expanding universe had something that Friedmann's earlier discovery did not: it could be tested with real-world observations.

Lemaître took his analysis a step further and looked for more solutions. To his surprise, he found that the static models that Einstein and de Sitter had been promoting were very special cases, almost aberrations of Einstein's theory of spacetime. While de Sitter's model could be recast as an evolving universe, Einstein's model suffered from an instability that could rapidly kick it off-kilter. If, in Einstein's universe, there was even the smallest degree of imbalance between matter and the cosmological constant, the universe would rapidly start to expand or contract, rolling away from the placid state that Einstein so desired. In fact, as Lemaître found, Einstein's and de Sitter's models were but two in a vast family of models, all of which expanded with time.

The de Sitter effect had not gone unnoticed among astronomers. In fact, in 1915, even before de Sitter first proposed his model and its hallmark signature, an American astronomer, Vesto Slipher, had measured the redshifts of smudges of light, known as nebulae, scattered throughout the sky. He achieved this by measuring the spectra of these nebulae. The individual elements that make up a light-emitting object, be it a light bulb, a hot piece of coal, a star, or a nebula, emit a unique pattern of wavelengths of light. When measured with a spectrometer, these wavelengths appear as a series of lines like a bar code. This bar code is known as an object's spectrum.

Slipher used his equipment at the Lowell Observatory in Flagstaff, Arizona, to measure the spectra of nebulae scattered all over the sky. He then compared his measured spectra with what he would have obtained if he had measured an object made of the same elements sitting on his desk in his office. (The spectra for the elements making up the nebulae were perfectly well known so he didn't actually need to repeat the experiment in his office.) He found that his measurements of the nebulae's spectra were all displaced relative to what he expected. The bar codes were shifted either to the left or to the right.

The shift in the spectra implied that the measured objects were in motion. When a source of light is moving away from an observer, the wavelengths in its spectrum appear to stretch. The net effect is that light will look redder. Conversely, if a source of light is moving toward the observer, its spectrum is shifted to shorter wavelengths and will look bluer. This effect, known as the Doppler effect, is something you have probably experienced in the context of sound. Imagine a speeding ambulance coming down the street toward you — the pitch of its siren changes as it passes by, shifting to a lower pitch as it moves away. This same effect in light enabled Slipher to figure out how things were moving in the universe.

Slipher's results weren't altogether surprising. He expected things to move around, buffeted by the gravitational pull of nearby objects. In fact, one of his first measurements seemed to indicate that one of the brighter nebulae, Andromeda, was moving closer to us: its light was blueshifted. But Slipher was systematic and recorded spectra of a few more nebulae. What he found was puzzling — almost all the nebulae seemed to be drifting away from us. There was a trend.

In 1924, a young Swedish astronomer named Knut Lundmark took Slipher's data and made a rough guess of how far away from us the different nebulae were. Lundmark still couldn't tell exactly how far away each nebula was and wasn't entirely sure about his results. But lying there in front of him was the telltale trend — the farther away the nebulae were, the quicker they seemed to move.

Now, in 1927, the Abbé Lemaître had rederived the trend that appeared in de Sitter's model and that Slipher seemed to see in the data. Indeed, his calculations predicted that measuring the redshifts *and* distances of faraway galaxies should reveal a linear relation between the two. Plotted on a graph, with distance on the horizontal axis and redshift on the vertical axis, the galaxies should all fall approximately on a straight line. Unaware of Friedmann's work, Lemaître wrote up his results for his doctorate and published them in an obscure Belgian journal. He included his calculations and a short section discussing the observational evidence, working out the slope of the linear relation that Eddington, Weyl, and he himself had found. The observational evidence for expansion was tentative and contained large errors, but it was tantalizing how everything seemed to fit together.

To Lemaître's utter dismay, his work was completely ignored by relativity's leading theorists, including Eddington, his former adviser. When Lemaître met Einstein at a conference later that year, Einstein was unimpressed by Lemaître's work. Einstein graciously pointed out to Lemaître that his work merely replicated Alexander Friedmann's findings. While Einstein had conceded that Friedmann's calculations were correct, he clung to his belief that these strange expanding solutions were a mathematical curiosity, unrepresentative of the real universe, which he knew to be static. He concluded his appraisal of

Lemaître's work with a dismissive zinger: "Although your calculations are correct, your physics is abominable." And with that, at least for a while, Lemaître's universe disappeared into the wilderness.

Edwin Hubble was much more respected for his problem-solving skills than for his charming personality. He had studied at the University of Chicago, where he had become a boxing champion, or so he claimed. Then he spent a few years as a Rhodes Scholar at the University of Oxford, picking up an infuriating faux English accent that would stick with him for the rest of his life. He complemented his pompous demeanor with a tweed suit and pipe, the embodiment of an English country squire. After Oxford, Hubble had fought in the Great War, like Friedmann and Lemaître, but had arrived just as the war ended.

In the late 1920s, people paid attention to Hubble's work because he had struck gold a few years before. At the beginning of the twentieth century it was well established that we live in a vast whirlpool of stars that make up our galaxy, the Milky Way. At the time, an unanswered question hung over astronomy: Was the Milky Way the only galaxy, a lonely island in the emptiness of space, or was it one of many galaxies in the cosmos? If you looked out at the night sky, among the stars and planets, there were faint, mysterious smudges of light, the same nebulae that Slipher had looked at and measured. Were these nebulae just developing stars in the Milky Way or distant other galaxies in the making? If the nebulae were indeed other galaxies, that meant that the Milky Way was only one galaxy among many.

Hubble answered that question by measuring the distance of one particular nebula, Andromeda. He had realized that he could use very bright stars known as Cepheids as beacons. By measuring how much dimmer the Cepheids he could see in Andromeda were compared to ones close by, he was able to figure out Andromeda's distance from Earth. The dimmer it looked, the farther away it had to be. The distance to Andromeda that Hubble came up with was enormous: almost a million light-years, five to ten times more than what was then the estimate of the size of the Milky Way. Andromeda couldn't be part of the Milky Way — it was too far away. The natural explanation was that Andromeda was simply another galaxy, just like the Milky Way. And if this was true of Andromeda, why shouldn't it be true of many other nebulae? With that one measurement, in 1925, Hubble made the universe a much bigger place.

In 1927, Hubble attended an International Astronomical Union meeting in Holland. He heard the fuss that was being made about de Sitter, Eddington, and Weyl's prediction for the redshift effect in the nebulae and learned how Slipher's measurements just might be the first hint that the effect was in the data. Lundmark's attempt to piece together a plot comparing velocities with distance showing a relation between the two had been published in 1924, just before Hubble's measurement of the distance to Andromeda, and his results had been met with skepticism. The Abbé Lemaître had used Hubble's distance measurements for his 1927 paper, but it had been published in an obscure Belgian journal, in French, and no one had read it. Hubble saw an opportunity to step in and detect the de Sitter effect himself, superseding all the previous attempts and positioning himself as the discoverer.

Hubble enlisted a member of the technical staff on Mount Wilson, Milton Humason. Night after night, Hubble had Humason set up the prisms at the telescope on Mount Wilson high in the mountains above Pasadena, California, and measure spectra. It was thankless work. The dome was cold and dark, and the iron floor left Humason's feet numb and sore. His back would ache from peering awkwardly through the eyepiece trying to find the spectral lines of his selection of nebulae. He knew that he had to do better than Slipher and look at really faint nebulae. The fainter they were, the farther away they would be. But he was battling with an instrument that wasn't really set up to do these kinds of measurements. It would take him two or three days to get a spectrum, while other telescopes could already do it in a few hours.

While Humason looked for redshifts, Hubble focused on determining distances. He measured the amount of light each nebula was emitting and compared the results. From this, he could get a rough idea of how far away the objects were, comparing to his measurement of the distance of Andromeda. He then combined his measurements of distance with Slipher's and Humason's measurements of redshifts to look for a linear relation between the two, the telltale sign of the de Sitter effect.

By January of 1929, Hubble and Humason had redshifts for forty-six nebulae. Of these, Hubble had distances for twenty-four, the closer ones for which Slipher had measured the redshifts. He plotted them on a graph: the x axis denoted the distances and the y axis denoted the apparent velocities determined by the observed redshifts. There was still a lot of scatter, but it looked better than Lundmark's or Lemaître's attempts, and there was a distinct trend: the farther away the nebulae were, the larger the redshift.

Hubble submitted for publication, without Humason, a short paper, "A Relation Between Distance and Radial Velocity Among Extragalactic Nebulae," plotting out his data. Lundmark had been there before him, but although Hubble mentioned Lundmark's work in passing, he hyped the importance of his own result. In his last paragraph he wrote, "The outstanding feature, however, is the possibility that the velocity-distance relation may represent the de Sitter effect, and hence the numerical data may be introduced into discussions of the general curvature of space." In a short, modest paper submitted on the same day, Humason published his measurements of redshift and distance to a nebula that was twice as distant as all the ones that Hubble had considered in his paper. It also seemed to lie along the redshift relation that Hubble was finding. There it was, the de Sitter effect.

Although Lundmark and Lemaître had been there before, Hubble's discovery of the linear relationship between redshift and distance was the catalyst that brought cosmology together. In the years that followed Hubble's seminal paper of 1929, the ideas of Einstein, de Sitter, Friedmann, and Lemaître, which had been fermenting during the previous decade or so, would finally be reconciled into one simple picture. And even though the evidence for the recession of galaxies was already sitting in Slipher's data and Lundmark's and Lemaître's tentative analyses, it was Hubble's and Humason's papers that convinced astronomers that the de Sitter effect might be real.

A year after Hubble's paper was submitted, Eddington wrote up a discussion of the de Sitter effect and Hubble's observations in *The Observatory*, the same journal that had published his pacifist pleas during the dark days of the Great War. The Abbé Lemaître, firmly ensconced at the University of Louvain, read Eddington's article and was non-plussed. There was no mention of his work — his far simpler model of an expanding universe had been forgotten. Lemaître immediately sent Eddington a letter, describing his work from 1927 in which he had shown that there were other solutions to Einstein's equations in which the universe expanded. At the end of his letter, he added, "I send you a few copies of the paper. Perhaps you may find occasion to send it to de Sitter. I sent him also at the time but probably he didn't read it." Eddington was mortified. His "brilliant" and "clear-sighted" student had kept him up-to-date with his forays in relativity, yet Eddington had simply dismissed and forgotten his work. He rapidly set to work promoting Lemaître's view of the universe and convincing de Sitter to drop his own model and adopt Lemaître's. Now it was Einstein's turn to be won over by the expanding universe.

Einstein's years in the limelight had distracted him from the tumultuous progress that was being made with his theory by Friedmann and Lemaître and the observations of receding galaxies. But in the summer of 1930 he too had to recognize that something was up. During a visit to Cambridge, where he stayed with Eddington and his sister, he was infected by Eddington's enthusiasm for Hubble's results and Lemaître's universe. On one of his many trips, he stopped in California and met Hubble at Mount Wilson, where they awkwardly discussed the new vision of the universe. Einstein had yet to become fluent in English and Hubble couldn't speak German, but together they saw how the expanding universe was being adopted by physicists and astronomers alike. And so, on another trip, now to Leiden, Einstein sat down with de Sitter and embraced the new cosmology that was emerging from his theory, proposing his own version of an expanding universe. The two agreed to drop the fix that Einstein had been compelled to add to make his theory work and give him a static universe. Out went the cosmological constant that Einstein had added as an afterthought in 1917.

After discovering the expanding universe in Einstein's equations, Lemaître wanted to take Einstein's general theory of relativity even further. He realized that Einstein's theory could say something about the beginning of time. Indeed, if you accept that the universe is expanding, the next obvious question is how and why it started to do so. If you follow the universe back in time, you come to a point where the whole of spacetime was squashed into a single point. It is a bizarre state of affairs, unlike anything we see in the natural world around us. Yet that is what Friedmann's and Lemaître's models seemed to show: an initial moment when spacetime comes into being.

So Lemaître proposed a completely radical idea for how the universe could have begun. It involved a true beginning to everything. In his view, the universe had emerged from a single thing: a primeval atom, or "primordial egg," as he liked to call it. This atom would have spawned the material that fills the universe today. The atom would have decayed according the laws of quantum physics that were just beginning to be understood, just like the radioactive decay of particles that had been observed in the laboratory. The progeny of the atom would themselves decay into more particles, and so on and so on.

It was a simple, speculative, almost biblical model, but Lemaître was at pains to keep religion out of his proposal. As a priest, he risked more than anyone being accused of bringing his faith into what was ultimately a purely scientific hypothesis. He published a short paper in *Nature* with the title "The Beginning of the World From the Point of View of the Quantum Theory." The title said everything. This wasn't divine intervention or a theological construct. It was the practical outcome of the cold, impartial laws of physics. Nature made it that way. He summarized his view thus: "If the world has begun with a single quantum, the notions of space and time would altogether fail to have any meaning at the beginning; they would only begin to have a sensible meaning when the original quantum had been divided into a sufficient number of quanta. If this suggestion is correct, the beginning of the world happened a little before the beginning of space and time."

In January of 1931, Eddington told the audience of his presidential address at the British Mathematical Association what he thought of Lemaître's newest idea, announcing, "The notion of a beginning of the present order of Nature is repugnant to me." Eddington had championed Lemaître's work on an expanding universe and had convinced Einstein to give up his static universe. Lemaître owed his international celebrity to Eddington. But this newest idea of Lemaître's was just too much for Eddington to stomach. It pushed Einstein's theory of space-time beyond its valid limits, or so Eddington thought, and he let everyone know.

Just as Einstein had dismissed the expansion of space in Friedmann's and Lemaître's work, Eddington refused to accept what the mathematics was telling him. Instead, he proposed another solution. With Hubble's and Humason's evidence for the recession of galaxies, Einstein's static universe had been cast away, but only just. Lemaître, in an attempt to explore all the possible solutions for the universe, had shown that Einstein's static universe had a catastrophic property that could work to Eddington's advantage — it was unstable. If you added just a little bit of stuff to Einstein's static universe, an extra galaxy, star, or even just an atom, it would start contracting to a point. Conversely, if you took any matter away, it would start expanding, ultimately behaving just like the universe that Friedmann and Lemaître had found. It was this instability that Eddington would retrofit to explain the expansion.

Eddington's proposal for how expansion would start was patchy and unfinished but believable and simple. The universe would begin just as Einstein had proposed: static and stagnant. In fact, it was a misnomer to say that the universe *began;* the universe could have been suspended in this state for an infinite amount of time until, as Eddington proposed, matter would somehow, in some yet-to-be-determined fashion, start to clump. The clumps would form stars and galaxies, and the empty space in between would play into the instability in Einstein's model and start to expand. A timeless universe would segue beautifully into an expanding universe.

While Eddington remained unconvinced by Lemaître's radical proposal for the beginning of the universe, Einstein thought differently. In the winter of 1933, both Einstein and Lemaître were traveling in the United States and converged on the balmy campus of the California Institute of Technology in Pasadena, where the abbé was asked to give two lectures. Their meeting in Solvay in 1927, during which Einstein had dismissed Lemaître's work and thrown it on the pile of correct but irrelevant consequences of his own theory, had not gone too well. But this time it was different, and Lemaître was respected as one of the leading lights in the new cosmology. During their sojourn there, the two men would wander through the gardens of the Athenaeum, the social heart of the Caltech faculty, engrossed in conversation. The *Los Angeles Times* would describe the two men as having "serious expressions on their faces indicating that they were debating the present state of cosmic affairs." It was fitting that Einstein was sitting through Lemaître's lectures at the place where the recession of galaxies had been discovered. At the end of one of Lemaître's seminars, he stood up and said, "This is the most beautiful and satisfactory explanation of creation to which I have ever listened."

After more than a decade of being misled by his own misguided intuition, Einstein finally saw the light. It was an interesting turn of events. The creator of the general theory of relativity hadn't been brave enough to accept the predictions that his theory made about the universe and had tried to fudge the answer by introducing a fix. It was only by embracing general relativity in its full mathematical glory that Friedmann and Lemaître had been able to propose an evolving, expanding universe, and the observational data had proved them right. Einstein's praise crowned Lemaître in the eyes of the popular press. Just as Einstein himself had been propelled into the limelight, Lemaître was now acclaimed as the "World's Leading Cosmologist." Lemaître would go on to become one of the grand old men of modern cosmology. Along with those of Alexander Friedmann, his ideas set the scene for the revolution in cosmology that would take place almost thirty years later.

Chapter 4

Collapsing Stars

R OBERT OPPENHEIMER wasn't particularly interested in the general theory of relativity. He believed in it, as any sensible physicist would, but he didn't think it was particularly relevant for physics at the time. Which makes it ironic that he would discover one of the strangest, most exotic predictions of Einstein's theory: the formation of black holes in nature.

Oppenheimer's interests lay in the *other* new theory that had taken off over the previous decade. He had cut his teeth and become famous as a quantum physicist, studying with the great and good of modern physics in Europe, and had eventually created the leading group in quantum physics in the United States, based at the University of California's Berkeley campus. To some extent, it was the rise of quantum physics and of men like Oppenheimer that was responsible for exiling Einstein's theory to a period of stagnation and isolation. Yet, in 1939, with his student Hartland Snyder, in trying to understand what would happen at the endpoint of the life cycle of heavy stars, Oppenheimer found a strange, incomprehensible solution to the general theory of relativity that had been lurking in the background for almost twenty-five years. Oppenheimer showed that if a star is big and dense enough, it will collapse out of sight. As he put it, after a while "the star tends to close itself off from any communication with a distant observer; only its gravitational field persists." It would seem as if a mysterious shroud had arisen around the collapsing ball of light and energy, hiding it from the outside world, and spacetime would wrap itself up in an impossibly tight knot. Nothing would be able to escape outside the shroud, not even light. Oppenheimer's result was yet another mathematical oddity that emerged from Einstein's equations, and many found it too difficult to stomach.

Almost a quarter of a century before Oppenheimer and Snyder found their result, the German astronomer Karl Schwarzschild had sent a letter to Einstein, signing off, "As you see, the war is kindly disposed toward me, allowing me, despite gunfire at a decidedly terrestrial distance, to take this walk into this your land of ideas." It was December of 1915 and Schwarzschild was writing from the trenches on the eastern front. He had volunteered immediately after the outbreak of the First World War in 1914, even though, as the director of the Potsdam Observatory, he was not required to fight. But, as Eddington later said of him, "Schwarzschild's bent was more practical." Like Friedmann, he had brought his ability as a physicist to bear on his military service, even submitting a paper to the Berlin Academy on "The Effect of Wind and Air-Density on the Path of a Projectile."

While in Russia, Schwarzschild received the latest copy of the *Proceedings of the Prussian Academy of Sciences*. In it he found Einstein's brief but breathtaking presentation of his new general theory of relativity. He had set to work unpicking the field equations that Einstein was proposing, looking at the simplest, most physically interesting situation he could think of. Unlike Alexander Friedmann and Georges Lemaître, who would years later look at the universe as a whole, Schwarzschild decided to focus on something less grand: the spacetime around a spherical mass such as a planet or a star.

When tackling a tangled mess of equations like the ones Einstein proposed, it helps to simplify. By looking at the spacetime around a star, Schwarzschild could focus on finding solutions that were static and didn't evolve with time. Furthermore, he wanted a solution that looked exactly the same at the pole as near the equator so that all that should matter was the distance of any point in space from the center of the star.

Schwarzschild's solution was immensely simple, a condensed formula that took almost no time write down. And to some extent, it was obvious. If you were located a fair distance away from the star's center, its gravitational field behaved much as Newton centuries before had predicted — the gravitational attraction of the star would depend on its mass and would fall as the square of the distance. Schwarzschild's formula was different, true, but the differences were very small — just enough to explain the drift in Mercury's orbit that had been hovering over Einstein's whole endeavor.

But as you moved closer to the star, something very strange happened. If the star was small but heavy enough, it would be shrouded by a spherical surface that kept everything behind it hidden from sight — the same surface that Oppenheimer and Snyder would find many years later. This surface had a devastating effect on anything that tried to pass through it. If anything flew too close to the star and fell within that spherical boundary, it would never be able to get out again — it was a point of no return. To get out of Schwarzschild's magic sphere, you would need to travel at speeds greater than the speed of light. And that, according to Einstein's theory, was impossible. Schwarzschild had discovered what would, more than half a century later, be called black holes.

Schwarzschild rapidly wrote up his results and sent them to Einstein in a letter, asking him to present them at the Prussian Academy of Sciences. Einstein approved and responded by saying, "I had not expected that one could formulate the exact solution of the problem in such a simple way." In late January of 1916, Einstein presented Schwarzschild's solution to the world.

Schwarzschild would never get to explore his solution any further, let alone hear of Oppenheimer and Snyder's calculation. For a few months later, while still in Russia, Schwarzschild contracted pemphigus, a virulent blistering autoimmune disease. His own body turned against him, and he died in May 1916.

Schwarzschild's solution was rapidly adopted by Einstein and his followers. It was simple, easy to work with, and perfect for making predictions. It could be used, for example, in a model of the sun to work out the motion of the planets and make an accurate prediction of the precession of Mercury's orbit. It also accurately predicted the bending of light that Eddington set out to find in Príncipe. Schwarzschild's solution served the new relativists well, except for that unfathomable property of the strange surface shrouding the center of certain dense, small stars and keeping everything out.

There was no denying the surface was there in the equations and the solution. It was a valid solution of Einstein's general theory of relativity. But did it actually exist in nature?

During the 1920s, Arthur Eddington turned to figuring out how stars form and evolve. He wanted to completely characterize the structure of stars using fundamental laws of physics couched in the correct mathematical equations. He wrote, "When we obtain by mathematical analysis an understanding of a result . . . we have obtained knowledge adapted to the fluid premises of a natural physical problem." With the mathematics in hand, it would simply be a matter of solving equations, just as with general relativity. In 1926 Eddington published a book, *The Internal Constitution of the Stars,* which rapidly became the bible for stellar astrophysics. Not only was Eddington a world authority on general relativity; he was also the leading light on stars.

Stars had until then been a bit of a mystery. For a start, no one had a clear idea of how they could emit such copious amounts of energy. It was Eddington who came up with a plausible mechanism for how stars are fueled. To understand his idea, we need to take a close look at the simplest atoms. A hydrogen atom is made up of two particles, a proton (which is positively charged) and an electron (which is negatively charged). The proton and electron are held together by electromagnetic force, which causes opposite charges to attract one another. The proton is approximately two thousand times heavier than the electron and so makes up almost all the weight of the hydrogen atom.

A helium atom consists of two electrons and two protons. But it also contains two *neutral* particles at its core: neutrons, which have almost exactly the same weight as protons. A simple model of the helium atom shows a nucleus made up of two protons and two neutrons orbited by the two electrons. Almost all the weight of the helium atom is made up of the four particles in the nucleus, and one would expect helium to be four times heavier than hydrogen. But helium is slightly lighter, by about 0.7 percent, than the expected mass of four hydrogen atoms. Some of its mass seems to be missing. And where there is missing mass, according to Einstein's special theory of relativity, there is missing energy. This was Eddington's cue.

Eddington argued that the interconversion between hydrogen and helium might be the source of energy in stars. Hydrogen nuclei would slam together in the intense, hot inferno at the core of stars. Some of the protons, through radioactive decay, would transform into neutrons, and collectively the protons and neutrons would form helium nuclei. In the process, each atom would release a minute amount of energy. The combined energy released by the atoms would be enough to fuel the star and emit light. If most of the sun started in the form of hydrogen, it should be able burn for almost 9 billion years before its conversion into helium is complete. Given that the Earth is currently about 4.5 billion years old, the numbers seemed to add up.

In his book, Eddington created a whole edifice for explaining stellar astrophysics. After proposing a source for stars' energy, he explained why they didn't collapse: they could withstand the pull of gravity by radiating all the energy they produced outward. Stars were perfect physical systems that could be described in terms of his equations. Yet *The Internal Constitution of the Stars* told an incomplete story. Eddington could describe the life of stars in terms of his mathematical pyrotechnics, but he stopped short of explaining their death. His own rationale led to the logical conclusion that at some point a star's fuel would run out and the radiation preventing it from collapsing under the force of its own gravity would disappear. As he said in his book, "It would seem that the star will be in an awkward predicament when its supply of subatomic energy ultimately fails. . . . It is a curious problem and one may make many fanciful suggestions as to what actually will happen." And, of course, one possible fanciful suggestion would be to embrace Einstein's theory and Schwarzschild's solution so that, as Eddington wrote, "the force of gravitation would be so great that light would be unable to escape from it, the rays falling back to the star like a stone to the earth." That was too far-fetched for Eddington, merely a mathematical result. For, as he declared, "when we *prove* a result without understanding it — when it drops unforeseen out of a maze of mathematical formulae — we have no grounds for hoping that it will apply."

Without being fanciful, then, what *would* happen when the fuel ran out? There were hints about the graveyard for stellar collapse in observations made in 1914. Astronomers had peered at the brightest star in the sky, Sirius, which is almost thirty times brighter than the sun, and observed an odd, dim companion orbiting it. Dubbed Sirius B, despite its dimness, it was incredibly hot and had remarkable properties: Sirius B had about the same mass as the sun, yet its radius was much smaller than Earth's. This meant that the companion star was very, very dense. By the early twenties, this object had been named a white dwarf and stood out as a mystery in the stellar zoo, a possible endpoint for the life cycle of a star. A key to explaining white dwarfs and their fate would come from the newfangled theory of quantum physics.

Quantum physics divided nature into its smallest constituents and put it back together in an outlandish way. It emerged from the bizarre phenomena that were being observed in the nineteenth century when physicists discovered that compounds and chemicals reemit or absorb light in a peculiar fashion. Rather than emitting or absorbing light in a continuous range of wavelengths, the substances would throw off light only in a discrete set of specific wavelengths, creating the bar-code-like spectra that would later reveal redshifting to Vesto Slipher and Milton Humason. The Newtonian physics that reigned at the time, allied with Maxwell's theory of electricity and light, couldn't explain this strange phenomenon.

During his miraculous year in 1905, Einstein had tackled another odd experimental fact: the photoelectric effect. If you bombard a metal with light, its atoms will soak up the light and sometimes release an electron. As the phenomenon's discoverer, Philipp Lenard, described it, "By mere exposure to ultraviolet light, metal plates give off negative electricity to the air." You might think that all you need to do is blast enough light at the metal for this to happen, but that isn't the case. Only if the beam has exactly the right energy and frequency will an electron be emitted. Einstein looked at this effect and conjectured that light comes in chunks of energy, quantized in the same way that matter breaks down into fundamental particles. Only when one of these light particles has just the right frequency would the photoelectric effect come into play. Einstein called these particles "light quanta" and they later became known as photons.

As experimental techniques advanced at the turn of the twentieth century, nature began to appear chunky and discrete, not smooth and continuous. In other words, nature seemed to be quantized. In the early twentieth century, a makeshift model for nature at the smallest scales began to emerge, a motley set of new rules for how atoms behaved and how they interacted with light. While Einstein himself occasionally made a contribution to this new science, he mostly observed the developments with some disbelief. The new rules proposed for a quantized world were clunky and didn't fit the elegant mathematical picture that had emerged from his principles of relativity.

In 1927, the rules of quantum physics finally fell into place. Two physicists, Werner Heisenberg and Erwin Schrödinger, each independently came up with new theories that could consistently explain the quantum nature of atoms. And like Einstein when constructing his general theory of relativity, the two men had to couch their versions of quantum theory in new mathematics. Heisenberg used matrices, tables of numbers that have to be worked with very carefully. Unlike ordinary numbers, if you multiply two matrices A and B, you will normally get a different result than you would if you were to multiply B with A, which can lead to some quite startling results. Schrödinger opted to describe reality — the atoms, nuclei, and electrons that stuff is made of — as matter waves, exotic objects that, just as in Heisenberg's theory, would lead to some strange physical phenomena.

The most notorious result to come out of the new quantum physics was the uncertainty principle. In classical Newtonian physics, objects move in a predictable way in response to outside forces. Once you know the exact position and velocities of a system's constituents and any forces acting on the system, you can predict all of the system's future configurations. Prediction becomes particularly easy; all you need to know is each particle's position in space and the direction and magnitude of its velocity. But in the new quantum theory it was *impossible* to know both the position and the velocity of a particle with perfect accuracy. A particularly persistent and stubborn experimenter in a lab who tries to pin down the position of a particle with perfect precision will have *absolutely no idea* what its velocity is. You could imagine that it is like working with an angry caged animal: the more you try to confine it, the more furious it will get, pounding on the walls of its cage. If you put it in too small a box, the pressure from its pounding on the walls will be immense. Quantum physics brought uncertainty and randomness into the heart of physics. It was precisely this randomness that could be put to use in solving the problem of white dwarfs.

Subrahmanyan Chandrasekhar yearned to do great things, almost desperately. Born into an affluent Brahmin family in India, Chandra, as he became widely known, was an intense and committed student. He excelled at mathematics and was meticulous and fearless at calculating. While studying at the University of Madras, he was exposed to the new ideas coming over from Europe, starstruck by the great men who were building the new physics of the twentieth century. From a young age, and with a feverish passion, he set about trying to join the fray of modern physics. As he said, later in life, "Certainly one of the earliest motives that I had was to show the world what an Indian could do."

Chandra was entranced by the new quantum physics. He read all the new textbooks that came his way, among them Eddington's recently published book, *The Internal Constitution of the Stars.* But what really won him over was a book on the quantum properties of matter by the German physicist Arnold Sommerfeld. Inspired by Sommerfeld's work, he set to work making a name for himself by writing papers on the statistical properties of quantum systems and how they interact. One of the first papers he wrote was published in the *Proceedings of the Royal Society* when Chandra was not yet eighteen years old. Clearly capable of taking part in the great discoveries of the new quantum physics in Europe, Chandra chose England to pursue his calling and set off on the long trip to Cambridge for his PhD.

It was during his long voyage on a ship of the Lloyd Triestino line that Chandra made the startling discovery that would transform his life. Obsessed with his work, he decided to spend his trip focusing on a paper written by Ralph Fowler, one of Eddington's colleagues at Cambridge, which seemed to solve the problem of white dwarfs. Fowler had invoked two quantum concepts and dragged them into astrophysics. The first was Heisenberg's uncertainty principle, the fact that you couldn't pin down a particle and at the same time determine its state of motion or velocity. The second concept was the exclusion principle, which states that two electrons (or protons) within an atom cannot be in exactly the same physical state — the exotic matter wave that Schrödinger had proposed as the fundamental quantum description of a particle — at the same time. Indeed, it is as if there is a fundamental, inexorable repulsion between them, preventing them from occupying that same state.

Fowler took the uncertainty and exclusion principles and set out to apply them to Sirius B. He reasoned that the material in a white dwarf such as Sirius B was so dense that he could think of it as a gas of electrons and protons being squeezed together. The electrons are so much lighter that they are allowed to roam more freely and jiggle about much more vigorously. The exclusion principle means that electrons have to be careful not to encroach on one another's space, and as the density builds up, each electron has less and less space to move in. As each electron is pinned down more and more, the uncertainty principle kicks in and the velocities and motions get higher and higher, forcing the electrons against each other. These fast-moving, jiggling electrons lead to an outward push, a *quantum* pressure inside the white dwarf, that can counteract the pull of gravity. In a certain state, the gravity exactly balances the quantum pressure and the white dwarf can sit placidly, hardly glowing but resisting a catastrophic fate. Fowler's explanation cleared up Eddington's problem. It seemed that stars could end up as white dwarfs. It closed the narrative of stellar evolution and solved the cliffhanger in *The Internal Constitution of the Stars* — or so it seemed.

Chandra took another look at Fowler's result and did something very simple. He put in the numbers he expected for the density of the electron gas in the white dwarfs. The number he came out with was immense but unsurprising, exactly as Fowler had claimed in his paper. What Fowler had failed to do was work out how large the velocities of the electrons would actually be. When Chandra did this simple calculation, he was shocked: the electrons would have to be zipping around close to the speed of light. And this is where Fowler's argument fell apart, for he had completely ignored the rules of special relativity that are so important when things start moving at the speed of light. Fowler made the mistake of assuming that the electrons inside the white dwarf could move as fast as they wanted, even if that meant they were zipping around *faster* than the speed of light.

Chandra set out to fix Fowler's mistake. He followed Fowler's reasoning all the way up until the electrons were moving close to the speed of light. If the white dwarf was too dense, and the particles were indeed moving close to or at the speed of light, he used Einstein's special theory of relativity, which posited that they couldn't move any faster. The result he obtained was intriguing. He found that if the white dwarf became too heavy, it would also become too dense and the electrons would be unable to sustain the gravitational pull. In other words, there was a maximum mass for a white dwarf. In his calculation, Chandra found that it couldn't be larger than about 90 percent of the mass of the sun. (Years later it would be shown that the correct value is more like 140 percent of the mass of the sun.) If a star ended its life as a white dwarf heavier than this maximum mass, it would be unable to support itself. Gravity would win out and inexorable collapse would ensue.

When he arrived in Cambridge, Chandra gave Eddington and Fowler a draft of his calculation, but they ignored it. There was something deeply unsettling about the instability, which would wreck the edifice Eddington had so promisingly put forward and to which Fowler had added, and so the Cambridge men kept their distance. Over a period of four years, Chandra perfected his argument, and his confidence in his result grew. In 1933 Chandra finished his PhD and, at age twenty-two, was made a fellow of Trinity College. By 1935 Chandra had finessed his calculation still further and was prepared to present his results at one of the monthly meetings of the Royal Astronomical Society.

On January 11, 1935, Chandra stood up in front of a crowd of distinguished astronomers at the Royal Astronomical Society, at Burlington House in London. Carefully and meticulously Chandra worked through his results, presenting the details of his nineteen-page paper, which was about to be published by the *Monthly Notices* of the society. He finished by saying, "A star of large mass cannot pass into the white dwarf stage, and one is left speculating on other possibilities." This strange result was there in the mathematics and physics that they all believed and had to be taken seriously. When Chandra finished, there was polite applause and a smattering of questions. It was done.

The president of the RAS then turned to Eddington and invited him to step up to the podium to talk on his own paper, "Relativistic Degeneracy." Eddington stood up to give his brief, fifteen-minute talk. He carefully went over Chandra's claim that his calculation scuppered Fowler's solution to the problem of white dwarfs. And then he summarily dismissed Chandra's watertight argument. To Eddington, Chandra's result was "a reductio ad absurdum of the relativistic degeneracy formula." In fact, he firmly believed that "various accidents may intervene to save the star," and furthermore, "I think there should be a law of nature to prevent a star from behaving in this absurd way!" Eddington's authority was such that Chandra's talk was immediately dismissed by most of the audience. If Eddington thought it was wrong, it *must* be wrong.

Chandra had come up against the mighty Eddington and lost. He was sabotaging Eddington's beautiful story of how stars lived and died, and Eddington didn't like it. If gravitational collapse overcame everything, Schwarzschild's strange solution would have to be faced head-on, with all its bizarre consequences. As Chandra himself said, many years later, "Now, that clearly shows that . . . Eddington realized that the existence of a limiting mass implies that black holes must occur in nature. But he did not accept that conclusion. . . . If he had accepted that, he would have been 40 years ahead of anybody else. In a way it is too bad."

Chandra returned to Cambridge devastated. His run-in with Eddington was to mark him for the rest of his life. A few years later he was invited to take up a post in the Yerkes Observatory in Chicago.

He stopped working on white dwarfs and shied away from thinking of what would happen if indeed their masses were too large. Would they lead to the inexorable formation of Schwarzschild's solution, or would something prevent that from happening along the way? Robert Oppenheimer would be the one to answer those questions.

J. Robert Oppenheimer was a child of the quantum. Brought up in an affluent New York family with van Goghs hanging on their walls, Oppenheimer had a gilded education, first studying at Harvard and then, in 1925, moving to Cambridge. Oppenheimer's Harvard mentor wrote in his letter of recommendation to Cambridge that Oppenheimer "was evidently much handicapped by his lack of familiarity with ordinary physical manipulations," although he added, "You will seldom find a more interesting betting proposition." Oppenheimer's sojourn in Cambridge was a disaster and short-lived. After a nervous breakdown during which he assaulted one of his colleagues and confessed to trying to poison another, Oppenheimer decided to leave and try his luck in Göttingen.

Göttingen, the land of David Hilbert, had embraced quantum physics, and Oppenheimer couldn't have been at a better place to take part in the new revolution. Over the next two years he wrote a series of papers with his supervisor, Max Born, that would indelibly imprint his name in the history of quantum physics. Indeed, the Born-Oppenheimer approximation is still taught in universities today and is part of the paraphernalia used to calculate the quantum behavior of molecules. Oppenheimer finished his PhD in 1927 and a few years later returned to the United States to take up a position at the University of California at Berkeley.

At Berkeley, Oppenheimer set up one of the beacons of theoretical physics in 1930s America. Oppie, as he was fondly called, seemed to be able to hold forth on any topic, from art and poetry to physics and sailing. Sharp and incredibly quick at picking up on difficult concepts, he hopped from project to project, intellectually raiding new fields and making quick contributions that, while not necessarily profound, were undoubtedly timely and clever. He was impatient and sometimes cruel if he didn't agree with or understand an argument, but Oppenheimer's sheer magnetism and energy made him a natural leader, and he excelled at supporting and inspiring his group. He slowly and surely recruited a coterie of brilliant and enthusiastic students and researchers with whom he would tackle many of the new problems that were being discussed in Europe. Wolfgang Pauli, noting that Oppenheimer in his enthusiasm had a habit of muttering, dubbed his group the "nim nim boys." Berkeley was Oppenheimer's Göttingen, his Copenhagen.

And then, after nearly ten years of focusing almost exclusively on the quantum, in 1938, Oppenheimer became intrigued by Einstein's general theory of relativity. Like Chandra, he approached the theory from the quantum end, looking at how the quantum effects of matter might play off against the gravitational implosion of space and time.

Every summer Oppenheimer would head down to Southern California with his crowd of students and researchers and take up residence at Caltech, in sunny Pasadena. There he could talk to not only the other physicists but also the astronomers who had followed Hubble's success and had witnessed Lemaître's lectures on the primeval atom at first hand. There, they still held a flame for general relativity. It was in Pasadena that Oppenheimer first read a paper by the Russian physicist Lev Davidovich Landau on what would happen if the cores of stars were purely made of a compact mess of neutrons.

Landau was one of the leading lights in Soviet physics, growing up during the Russian Revolution, a truly brilliant physicist who benefited from the wave of modernization sweeping through the new Russia. Like Oppenheimer, he had spent time abroad, studying in the great laboratories of Europe and witnessing the birth of quantum physics. At nineteen he had already written a paper applying the new physics to the behavior of atoms and molecules, and when he returned to Leningrad, at twenty-three, he had earned the admiration of his older colleagues and was rapidly embraced by the Soviet system.

With his flair for solving difficult and complex physical systems with quantum physics, Landau had decided to look at a novel source for energy in stars: neutrons, the neutrally charged particles found in the nuclei of atoms. Over the previous decade, it had become clear that adding neutrons or protons to or removing them from nuclei could lead to a copious amount of *nuclear* energy. So Landau conjectured that if the cores of stars could be packed with neutrons, it might be possible to unleash enough nuclear energy to generate light. If the neutrons were packed together to a density that resembled that of the nucleus of an atom, they might just be the necessary fuel. This nuclear material would be impossibly heavy — a teaspoon of material would weigh tons. If an atom in the stars' bulk fell into the core, it would be smashed to smithereens, partly absorbed, and partly released as radiation. According to Landau, the neutron core fueled a star's brightness — it was what made the sun shine. Landau proceeded to work out how big the core had to be and that for such a core to be stable, it just had to be heavier than a thousandth of the weight of the sun. These cores could be tucked away at the center of stars, burning away and fueling starlight.

But as Landau was writing up his idea, he was also getting caught up in the wave of political repression that was sweeping the country. Two months after Landau published his short paper on neutron cores, "Origin of Stellar Energy," in *Nature,* he was arrested by the NKVD. He had been caught editing an anti-Stalinist pamphlet to be distributed at the 1938 May Day parade in Moscow in which Stalin was accused of being a Fascist "with his rabid hatred of genuine Socialism" who had "become like Hitler and Mussolini." Landau was incarcerated for a year in the Lubyanka prison, just after his *Nature* paper was feted in *Izvestia,* one of the main Soviet newspapers, as a source of pride for Soviet physics.

Oppenheimer was intrigued by the brevity of Landau's paper and the simple idea being proposed, so he decided to redo Landau's calculations himself. It took three collaborations with three gifted students, but he eventually got where he wanted to go. His first collaborator was Robert Serber. Together, they gently pulled apart Landau's idea that the neutron core could be easily tucked away in the sun, shrouded by the hot gases that puff the stars up, and showed that it was wrong. Oppenheimer and Serber published their letter, almost as short as Landau's, in October 1938 in the *Physical Review,* while Landau languished in the Lubyanka. Oppenheimer then took the next step with another student, George Volkoff. The pair studied the stability of neutron cores. Their calculation, published in January 1939, is a beautiful mix of mathematics using clever simplifications of Einstein's theory, with insightful physical intuition and hard calculations. They showed that neutron cores were incredibly unstable configurations and hence couldn't even be invoked to fuel the energy of very large stars, yet another blow for Landau's idea.

At the end of their paper, Oppenheimer and Volkoff pointed out that "a consideration of non-static solutions must be essential" to understand the long-term fate of the neutron cores. Then Oppenheimer set off to do the last piece with yet another student, Hartland Snyder, this time taking general relativity far beyond what anyone had ever attempted. Oppenheimer and Snyder calculated how space and time (and the neutron core) would evolve once the neutron star became unstable. To do so they used a clever idea to understand the results that they were getting: they placed a fictitious observer very far away from the implosion and another fictitious observer right on the surface of the neutron core and compared what those observers would see. They found that the two observers would see very different things.

A distant observer would see the neutron core implode. But as the surface of the neutron core got closer and closer to the strange shroud that Schwarzschild had found, the collapse would seem to proceed more and more slowly. At some point the implosion would be so slow that it would almost have ground to a halt. The wavelength of any light beam trying to escape from the neutron core would be stretched, redshifting more and more the closer the surface of the neutron core contracted to the critical surface. It would be as if space and time had stopped evolving, and the star would cease to communicate with the outside world. It was very similar to what Eddington himself had said more than a decade before in his book *The Internal Constitution of the Stars:* "The mass would produce so much curvature . . . that space would close up round the star, leaving us outside (i.e. nowhere)."

An observer riding the surface of the star as it imploded would see something completely different. He or she would witness the inexorable collapse of the neutron core, see the surface of the neutron core actually *cross* the critical radius and fall into the inner region of Schwarzschild's magic surface. And furthermore, this poor, doomed observer would see the formation of the dreaded surface that Schwarzschild had found, the point of no return from which nothing could exit. In other words, if you could sit at the right (or wrong) place, you could see the actual formation of Schwarzschild's solution.

Oppenheimer and Snyder had completed Eddington's life story of stars by showing that, indeed, if they were massive enough, they would collapse to form Schwarzschild's strange solution. It meant that Schwarzschild's solution might not be just an interesting, exotic solution to the general theory of relativity. These strange objects might actually exist in nature and had to be included in astrophysics, just like the study of stars, planets, and comets. Once again, general relativity had potentially revealed something unexpected and wonderful about the universe.

Oppenheimer and Snyder's paper was published on September 1, 1939, in the *Physical Review,* on the day Nazi troops marched across the Polish border. In the exact same issue was another paper, this one by a Danish physicist named Niels Bohr and his young American collaborator, John Archibald Wheeler. While they were also interested in neutrons and how they interact in extreme situations, the topic of "The Mechanism of Nuclear Fission" was completely different. Bohr and Wheeler were interested in modeling the structure of very heavy nuclei, such as those of uranium and its isotopes. If they could get this right, it might be possible to figure out how to extract the enormous amounts of energy locked up inside.

Throughout the 1930s, the zoo of atomic nuclei had begun to be understood in ever-increasing detail. Eddington had proposed that hydrogen nuclei could fuse together to form helium in the cores of stars, fueling starlight. This is known as nuclear fusion. On the other end of the range, it was believed that very heavy nuclei could be split into smaller nuclei, also releasing energy — in this case the process is known as nuclear fission. A question that was on everyone's mind was how to make nuclear fission efficient. Would it be possible to trigger nuclear fission in a clump of heavy atoms with a small amount of energy so that as each individual atom split, it would trigger yet another split? In other words, was it possible to trigger a chain reaction?

Bohr and Wheeler's paper pointed the way to nuclear fission and helped other physicists understand why uranium-235 and pluto-nium-239 might be the elements of choice to work, the sweet spot in the periodic table where fission might actually be easier to accomplish. Nuclear fission would dominate physics during the years that followed, eclipsing almost all other fields. An army of brilliant scientists turned their intellects to trying to understand how to harness fission, and Robert Oppenheimer was among them.

Oppenheimer, during his stay at Berkeley, had built a stunning group of young researchers and students who were willing to tackle any problem. He had developed a formidable reputation as an orga-nizer and group leader and would deploy his leadership skill to mar-shal his team to solve problems that were of interest to him. His col-leagues at Berkeley were beginning to synthesize the heavier, unstable elements in the cyclotron up on the Berkeley Hills. In 1941, one of his colleagues, Glenn Seaborg, discovered plutonium, opening one of the routes to fission. Oppenheimer was being caught up in the whirlwind of events and discoveries that characterized the development of nu-clear physics during the Second World War.

Oppenheimer was also outraged. The reported treatment of Jews in Germany and the diaspora of brilliant scientists fleeing Nazi oppres-sion who were washing up on American shores shocked him. As he developed his group at Berkeley, he also started to look around him, tentatively engaging with the teeming intellectual activity of the influx of European refugees. Although he refrained from being too active politically, he began paying attention. And with the onset of the war, nuclear fission became one of Oppenheimer's main concerns.

In 1942, Oppenheimer was asked to lead a task force of physicists based in Los Alamos, New Mexico, whose sole purpose was to pro-duce and control a chain reaction of nuclear fission. The task force included a host of young and not-so-young brilliant minds, from John von Neumann, Hans Bethe, and Edward Teller to the young Richard Feynman. The Manhattan Project focused its resources on producing the first atomic bomb, and in just under three years they had achieved their goal. When the two atomic bombs, "Little Boy" and "Fat Man," were dropped on Hiroshima and Nagasaki in August of 1945, around two hundred thousand people were killed. These devastating consequences were a harrowing testament to Oppenheimer's ability to harness the nuclear force in such a short period of time. With the success of the atomic bomb, the quantum firmly took center stage in the world of physics.

With so much attention focused on the war and the nuclear project, Oppenheimer and Snyder's seminal paper on black holes was kicked into the long grass, to be ignored and forgotten for years to come. What could have been the auspicious birth of one of general relativity's greatest concepts was indefinitely put off. The two grand old men of general relativity, Albert Einstein and Arthur Eddington, did nothing to save Oppenheimer and Snyder's finding from obscurity.

Eddington continued to insist that Chandra's calculation was wrong and misguided and that white dwarfs were the quiet endpoint of stellar evolution for stars of any mass. The continued unfettered collapse of a star until "gravity becomes strong enough to hold in the radiation" was simply absurd. Chandra recalled, almost half a century later, "For my part I shall only say that I find it hard to understand why Eddington, who was one of the earliest and staunchest supporters of the general theory of relativity, should have found the conclusion that black holes may form during the natural course of evolution of the stars, so unacceptable."

Einstein himself continued resisting the idea that the extreme form of Schwarzschild's solution — black holes — had any place in the natural world. He reacted in much the same way as he had to Friedmann and Lemaître's proposal of an expanding universe: it was beautiful mathematics but abominable physics all over again. After more than twenty years dismissing the more outlandish features of Schwarzschild's solution, he finally sat down and tried to come up with a reasoned argument for why they were of no physical significance in nature. In 1939, the same year Oppenheimer and Snyder devoted to determining the consequences of gravitational collapse, Einstein published a paper in which he worked out how a swarm of particles would behave as they collapsed through gravity. He argued that particles would never fall too close to the critical radius. He was too stubborn, setting up the problem in such a way that he got the answer he wanted: no black holes. He was wrong, once again, and just like Eddington he missed an opportunity to explore the full glory of his general theory of relativity.

Almost everyone's attention was elsewhere now, enthralled by the triumph of quantum physics. Most of the talented young physicists were focusing their efforts on pushing the quantum theory further, looking for more spectacular discoveries and applications. Einstein's general theory of relativity, with all its odd predictions and exotic results, had been elbowed out of the way and sentenced to a trek in the wilderness.

Chapter 5

Completely Cuckoo

DURING HIS FINAL YEARS, Albert Einstein lived a simple life. He would wake up late in his white clapboard house on Mercer Street near the heart of Princeton, New Jersey, where he lived with his sister, Maja. (His wife, Elsa, died in 1936, shortly after his arrival.) During the week, he would walk to Fuld Hall at the Institute for Advanced Study, where he had been based since 1933. Over the years he had become a familiar presence on the Princeton campus. Yet while he was more famous than ever before, he cut a lonely figure.

Einstein had been recruited to be one of the first permanent members of the institute, a privately funded haven for brilliant minds that had been set up by the Bamberger family. Einstein was surrounded by illustrious colleagues. There was John von Neumann, a mathematician who had worked on the atomic bomb and was one of the inventors of modern computers, and for a while the mathematician Hermann Weyl, one of David Hilbert's protégés, who had been one of the first to take up the banner of Einstein's theory of spacetime. Then there was Kurt Gödel, the philosopher and logician who wreaked havoc in twentieth-century philosophy with his incompleteness theorem. And of course there was Robert Oppenheimer, who had become the director of the institute in 1947. In the corridors Einstein might encounter distinguished visitors, architects of the quantum or of modern mathematics. But mostly he would retreat to his office.

After a few hours, Einstein would head back home for lunch and a nap. He would then wander over to his study and sit in his favorite chair with a rug around his legs, calculating, writing, and dealing with the multitude of letters that encroached on his life from the outside world. Letters from heads of state and dignitaries were interspersed with requests from aspiring young scientists and fans. At the end of the day, he would have an early supper, then listen to the radio and read for a bit before going to bed.

It was an unusually quiet life for a man who had reached such colossal fame. He wasn't forgotten. His name was just as recognizable to the public as Charlie Chaplin or Marilyn Monroe. He was a member of countless learned societies and had been awarded the keys to many cities. The cover of *Time* magazine featuring his picture became one of the iconic images of the new technological era. Every now and then, celebrities would still make their way to his door for a few hours with the great man. Jawaharlal Nehru and his daughter, Indira Gandhi, stopped by, as did the premier of Israel, David Ben-Gurion. The Juilliard String Quartet once came to play an impromptu concert in his front room.

Despite his global fame, Einstein kept mostly to himself. While he had a few younger assistants working with him, he chose to spend his time working alone. His general theory of relativity was still his pride and joy, and he would every now and then delve into it, moving beyond the solutions of Friedmann, Lemaître, and Schwarzschild and trying to find new, more complicated, but possibly more realistic ones. General relativity still had so much to give, but not many people were spending time on it, preferring instead to invest their efforts in quantum theory. Even Einstein himself chose to spend most of his time on a new, more ambitious theory that had been consuming him for almost three decades. And he would be shunned for it.

The Einstein of the 1950s could not be more different from the Einstein of the 1920s. Following his early scientific successes, Einstein had traveled the world, being treated like royalty, giving public lectures, debating other physicists, resisting and then embracing the discovery of the expanding universe. He was rewarded with the construction of the Einstein Tower on the outskirts of Berlin, in Potsdam, where observational research into his theory could be carried out. He was lauded at international meetings, where he was invited to opine on the newest developments in physics.

He had also seen the crescendo of anti-Semitic feeling in his homeland and, as the 1930s arrived, had felt the hard realities of the rise of the Nazi Party and its followers. His travel became more constricted, the death threats started to multiply, and even though his fame continued to grow, he became more wary of traveling through Europe to fulfill his many engagements.

Although he was somewhat shielded from the turmoil around him, a national treasure spared from the ugliness, Einstein had felt the dark underbelly of anti-Semitism early on. Shortly after his discovery of general relativity, a band of scientists, officially known as the Working Party of German Scientists for the Preservation of a Pure Science, took it upon themselves to campaign against his new theory. The Working Party smeared relativity as an example of "mass delusion" and attempted to build a case of plagiarism against Einstein. The movement recruited a world-renowned scientist as a vocal opponent to relativity: Philipp Lenard.

The Hungarian-born Philipp Lenard had won the Nobel Prize in 1905 for his work on cathode rays, and his experimental work had been at the heart of Einstein's early work on light quanta. His relationship with Einstein had been courteous throughout the lead-up to the discovery of general relativity. But he violently objected to Einstein's relativity — it was far too obscure and went against what he considered the "common sense" of any physicist. Lenard proceeded to write articles, dismissing Einstein's theory in the *Yearbook,* the same journal where, in 1907, Einstein had first presented the ideas that would lead to his general principle of relativity. A war of words ensued, in which Einstein dismissed Lenard as an experimentalist, not particularly capable of understanding his ideas. Lenard took offense, demanding a public apology. The public affray reflected badly on Einstein as well as Lenard and the "anti-relativists."

By 1933, Einstein had had enough of Germany. When the Nazi Party came to power, he decided to cut his ties with Berlin. He left Germany as it was entering its darkest days, and his theory became a target for the *Deutsche Physik,* or German Physics, movement. With the rise of the Nazi Party, Philipp Lenard's case, now with the vociferous support of another physicist and Nobel Prize winner, Johannes Stark, was much easier to make. According to Lenard and Stark, Einstein's theory was simply part of something insidious that was poisoning German culture: *Jewish physics.* In line with the grand plans of Nazi ideology, Jewish physics had to be eradicated from the system.

The years following Einstein's departure saw the systematic destruction of physics in the scientific community in Germany, which had been responsible for most of the greatest developments of the early twentieth century. By the time the Second World War broke out, all Jewish professors of physics had been removed from their university positions. Some of the most visionary thinkers in modern physics, instrumental in the creation of the new quantum physics, such as Erwin Schrödinger and Max Born, abandoned Germany. Some of them ended up contributing to the Allied atom bomb projects during the Second World War.

With the physics community seriously crippled, Johannes Stark set about establishing himself as the leader of the new Aryan physics. One of the fathers of the modern quantum theory, Werner Heisenberg, stood in his way. Heisenberg wasn't Jewish, but this didn't stop Stark. He wrote a piece for the official magazine of the SS labeling Heisenberg a "White Jew," as much a part of the decay of German science as all the others who had been ousted. Yet, surprisingly, Stark failed. Heisenberg had been at school with Heinrich Himmler, the commander of the SS. Himmler protected Heisenberg from further vilification. Indeed, Heisenberg ultimately ended up running the German atom bomb project, much to the consternation of his colleagues who had fled Hitler's Germany.

Einstein's departure left work on his theory in Germany in the doldrums. He had been lauded as a national hero during the Weimar Republic, but he rapidly disappeared from German culture during the Nazi years. Some of the ideas that had led up to the formulation of his special theory of relativity were included in textbooks, but the main physics textbook, Grimshels's *Lehrbuch der Physik,* made no mention of his name. Only after the war would Einstein's general theory of relativity be taken up again in Germany.

It wasn't only in Germany that Einstein's ideas were taking a battering. On the opposite side of the political spectrum, in the Soviet Union, relativity and quantum mechanics had occasionally run into trouble with the officially adopted philosophy, dialectical materialism, an integral part of Marxism. Based on the ideas of the German philosophers Friedrich Hegel and Ludwig Feuerbach, dialectical materialism was developed by Karl Marx in the mid- to late nineteenth century and was further refined by Friedrich Engels and numerous followers, notably Vladimir Lenin. In his 1938 article "Dialectical and Historical Materialism," Joseph Stalin defined, explained, and effectively canonized dialectical materialism as part of the official Soviet ideology. In this philosophy, the basis of everything was matter, and everything else emerged from that. Reality was defined by the way the material world behaved and was interrelated, preceding any form of thought and idealization. As Marx stated in his magnum opus, *Das Kapital,* "The ideal world is nothing else than the material world reflected by the human mind, and translated into forms of thought."

A practitioner of Marx's philosophy strove to explain everything in terms of the different constituents of the natural world and their interactions. Everything in the natural world contributed to a universe that was in a constant state of flux and evolution, punctuated by the most dramatic transformations that could arise from the gradual accumulation of the smallest changes. Crucially, the existence and evolution of matter were viewed as an objective reality whose laws were independent of observers and interpretations. Human knowledge was capable of approximating this objective reality faithfully and closely in a series of converging iterations, but the process would never be exhaustively complete and would never come to an end.

Most if not all physicists in the world would have no problem with materialistic views per se, and in fact in their work they all were practicing materialists without bothering to call themselves such. But the same physicists would definitely view with disdain and vehemently oppose any attempt by the philosophers to teach them how to do their research using the "correct methodology" advocated by a particular philosophical school. Marxism-Leninism was not just a particular philosophical concept; it was a powerful, all-reaching doctrine fully supported by the Soviet state. In the tense political atmosphere of the 1930s, 1940s, and 1950s, philosophical debates on the interpretation of quantum mechanics or relativity had the potential of deteriorating into accusations of disloyalty, sometimes with dangerous consequences.

Admittedly, the relativistic physics of Einstein as well as the emerging new radical ideas on the quantum, with their complexity and endless and often vague philosophical musings, were easy prey for Soviet philosophers of science. There was much that could be attacked in Einstein's theory of spacetime as well. First and foremost, it was the ultimate example of idealization. It had arisen from Einstein's now-famous thought experiments, with little or no input from the tangible, natural world. Furthermore, it was couched in the most abstruse mathematical language, a set of rules and principles that obscured interpretation, especially by the people who, like many philosophers, were not experts in sophisticated mathematics. Finally, to crown it all, Einstein's theory gave rise to an absurd universe with a defined origin, too close to the religious viewpoint that Soviet thought was so keen on eradicating from society. It didn't help that one of the lead contributors was a priest, the Abbé Lemaître, another corrupt foreigner from a decadent bourgeois society in its final throes. In fact, in a fierce rejection of non-Soviet thought, it was conveniently forgotten that the expanding universe had in fact first been proposed by the brilliant Russian and *Soviet* physicist, Alexander Friedmann. The debate smoldered for years, flaring occasionally, yet it would be too simplistic to view it as an ideological battle between brilliant physicists and ignorant orthodox philosophers. A number of physicists and mathematicians, some of them well known, joined the philosophers' ranks, and the dispute was severely aggravated by group allegiances and other factors not related to the subject of the discussion.

In 1952, Alexander Maximow, an influential Soviet philosopher and historian of science, published an article titled "Against the Reactionary Einsteinianism in Physics." Although the article was published in the obscure Soviet Arctic Navy newspaper, *The Red Fleet,* the reaction of physicists was strong: Vladimir Fock, a student of Friedmann and the leading Soviet relativist of that time, countered with his own article, "Against the Ignorant Criticism of Modern Theories in Physics." Before publishing the article, Fock, Lev Davidovich Landau, and other physicists appealed to the Soviet political leadership for support. In a private letter addressed to Lavrentiy Beria, a close associate of Stalin and the head of the Soviet nuclear and thermonuclear projects, they complained about the "abnormal state of affairs in Soviet physics," citing Maximow's article as an example of aggressive ignorance hindering the progress of Soviet science. The article was published, and Fock announced he had the government's support in the matter. Outraged, Maximow complained to Beria, insisting on his views, but by 1954, the group of Fock and Landau had prevailed. Of course, the highest political leadership of the Soviet Union had more urgent things to do than analyze the intricacies of Einstein's theories. Furthermore, Landau and the others had a very powerful argument on their side: they had worked on and delivered the Soviet atom bomb, and thus the theories on which their work was based, philosophical interpretation notwithstanding, were correct. By the mid-1950s, the ideological wars between Soviet philosophers and physicists had ended, and relativists were left alone. One of the last recorded vestiges of that battle was a 1956 note to the Central Committee of the Communist Party complaining about an "ideologically incorrect" plenary talk on the theory of an expanding universe by Evgeny Lifshitz, Landau's coauthor of the world-renowned *Course of Theoretical Physics.* The note was duly considered by the Central Committee, with no consequences.

The wars with Marxist philosophers had no bearing on the political repressions of 1937–38 and other years during which several extraordinarily talented Soviet physicists such as Matvei Bronstein, Lev Shubnikov, Semen Shubin, and Aleksander Witt perished while others were arrested, imprisoned, or exiled. Yet, while it seems that ideological battles had little if any influence on the course of the development of Einstein's relativity in the USSR, progress was slow, similar to what was happening in the West, due to the rapid rise of interest in quantum theory, the country's struggle for survival during rapid industrialization, the epic and victorious battle with European Fascism, and the subsequent nuclear race during the Cold War.

If the Soviet philosophers didn't approve of the mathematical idealism that had gone into the general theory of relativity, they certainly rejected Einstein's later work, for by the time Einstein arrived in Princeton, he had become obsessed with finding a grand unified theory. His general theory of relativity was still dear to his heart, but he wanted to do something bigger and better. He wanted to subsume general relativity into a theory that could bring *all* of fundamental physics into one simple framework. Einstein hoped to show how not only gravity but also electricity and magnetism, and possibly even some of the strange effects that were attributed to the quantum, could arise as the geometry of spacetime. But unlike his journey to general relativity, with his simple physical insights elegantly brought together with Riemannian geometry, Einstein approached his new challenge in a completely different way. He gave up on his formidable physical intuition to follow the math.

Einstein didn't come up with only one grand unified theory. For over thirty years he stumbled from theory to theory, sometimes discarding one possibility to pick it up again years later. One of his attempts extended spacetime into five dimensions instead of four. The additional spatial dimension was wrapped up and almost invisible. Its geometry, or curvature, would play the role of the electromagnetic field, responding to charge and currents exactly as James Clerk Maxwell had proposed in the mid-nineteenth century.

The idea of a five-dimensional universe wasn't originally Einstein's. It came from two young scientists, Theodor Kaluza, a lowly *privatdozent* in mathematics at the University of Königsberg, and Oskar Klein, a young Swedish physicist who had worked under Niels Bohr. Together they had worked out in detail how these five-dimensional spacetimes could mimic electromagnetism almost perfectly. The universes of Kaluza and Klein on which Einstein spent almost twenty years of his life are littered with a strange form of matter, an infinite variety of particles with a wide range of masses that should be all around us, warping the remaining geometry of spacetime. Einstein hoped, but was never able to show, that these extra fields might be inextricably tied to the quantum wave functions that Schrödinger had concocted in his quantum physics. Einstein gave up on these theories in the late 1930s, but interestingly enough the Kaluza-Klein theories would return in the 1970s when the idea of a unified theory took firm hold in theoretical physics.

Einstein devoted much more time to another theory for bringing together gravity and electromagnetism. He took his geometric framework for general relativity, the language that Riemann had proposed many decades before, and loosened it up. The original theory describing the geometry and dynamics of spacetime used ten unknown functions that had to be determined from his field equations. The fact that there were so many unknown functions, and that they were tangled up with each other in his original field equations, was one of the main reasons why general relativity was so hard to work with. But in his new theory, Einstein wanted to extend things by adding another six functions, three of which would describe the electrical part and another three of which would describe the magnetic part. The difficulty was how to bring these *sixteen* functions together in such a way that his theory would still be perfectly well defined and predictable. If he succeeded, the result, just like general relativity, should lead to the remarkable results that come out of *both* general relativity and electromagnetism. He wanted it to be mathematically beautiful, yet for decades he couldn't figure out how to make it so.

Einstein was onto something — the quest for a grand unified theory would come to dominate the physics of the late twentieth century — but during his lifetime he pursued this impossible quest alone. While he cut a solitary figure, working at the coal face of his new and fiendishly difficult theory, the outside world looked on with fascination. Every now and then Einstein would make the front page of one of the main newspapers. In November 1928, a *New York Times* headline hailed, "Einstein on Verge of Great Discovery," and a few months later, following a brief interview with Einstein, it reported, "Einstein Is Amazed at Stir Over Theory: Holds 100 Journalists at Bay for a Week."

The level of attention and excited anticipation lasted throughout the next quarter of a century. In 1949, the *New York Times* again declared, "New Einstein Theory Gives a Master Key to the Universe," and a few years later, in 1953, it trumpeted, "Einstein Offers New Theory to Unify the Law of the Cosmos." Despite all this attention in the popular media, among his colleagues Einstein had become somewhat irrelevant and his attempts at unification were widely dismissed.

While Einstein had escaped the torrent of abuse that was being leveled at his work in Germany, he found that general relativity was also disappearing from view in his new home, the United States. Around him, the bright young scientists with the potential to push general relativity ahead were being sucked up into the theory of quantum physics, teasing out its application to fundamental particles and forces.

In some sense it was understandable. General relativity had delivered a few great successes early on, such as the precession of the perihelion of Mercury and the gravitational bending of light. And it had led to the discovery of an expanding universe, a spectacular change in our worldview. But that was it. From then on, it seemed that it could only serve up somewhat unbelievable, *mathematical* results, like Schwarschild's or Oppenheimer and Snyder's solutions for a collapsing or collapsed star. There was a case for these bizarre solutions existing out there, in space, but no one had seen them, so they really had to be considered mathematical exotica. Quantum physics could be tested in the laboratory and could be used to build things. But it was clear that there were more strange things to be found in general relativity, as the logician Kurt Gödel was able to show.

Einstein wasn't always alone on his treks from his house to the institute. Often, this eccentric, rumpled-looking professor, with his straggly hair and kind gaze, would be accompanied by a small figure, always wrapped in a heavy overcoat, his eyes hidden by thick Coke-bottle glasses. While Einstein trundled distractedly up toward Fuld Hall, the other man would trail beside him, quietly listening to Einstein's monologues, responding in a high-pitched voice. Einstein relished these walks with this odd little man, who had been at the institute for as long as he had and confided in him. His friend was Kurt Gödel, the man responsible for dismantling modern mathematics. To Einstein's disbelief, Gödel would also poke a significant hole in his general theory of relativity.

Gödel had come out of the intellectual powerhouse that was Vienna in the early twentieth century. A culture of debate and modernity thrived in the coffeehouses of Vienna, which was home to Ernst Mach, Ludwig Boltzmann, Rudolf Carnap, Gustav Klimt, and any number of brilliant thinkers. The most prestigious of all informal meetings was the world-famous Vienna Circle. To belong to the Vienna Circle, you had to be invited. Gödel was one of the chosen few.

Unlike Einstein, Gödel had flown through his childhood education, obtaining perfect scores in all the subjects he was set and barreling through university, an outstanding student. He had flirted with physics but, unlike Einstein, had been drawn to how mathematics could be brought together into one logical framework. He rapidly mastered the developments that were coming fast and frequently from philosophers and mathematicians alike in their attempts to construct an ironclad theory of mathematics, impervious to irrationality, guesses, and tricks. Such was the plan set forward by David Hilbert, who reigned over mathematics in Göttingen.

David Hilbert firmly believed that all of mathematics could be constructed from a handful of statements, or axioms. With a careful and systematic application of the rules of logic, it should be possible to deduce *every single mathematical fact in the universe* from no more than half a dozen axioms. Nothing would be left out. The verification of any mathematical fact, from $2 + 2 = 4$ to Fermat's Last Theorem, should come from logical proof. Hilbert's program was the driving force behind mathematics when Gödel turned his sights on it.

While Gödel immersed himself in Vienna life, quietly attending the meetings of the Vienna Circle and watching the endless debates between the logicians and mathematicians on how to extend Hilbert's program to the whole of nature, he slowly and steadily chipped away at its fundamental premise. Then, in one fell swoop, he completely demolished Hilbert's plans with his own incompleteness theorem.

The incompleteness theorem states something incredibly simple. Whenever you describe a system mathematically, you begin with a set of axioms and rules. Whatever those initial statements are, Gödel showed that there will always be things that you can't deduce from them: true statements that you are unable to prove. If you stumble across a truth you can't prove using your axioms and your rules of logic, you can add it to your set of axioms. But Gödel's theorem showed that there will in fact always be an infinite number of these unprovable true statements. As you meander along picking up truths that you can't prove and adding them to your axioms, your simple, elegant deductive system becomes bloated, gigantic, and yet always incomplete.

Gödel's theorem torpedoed Hilbert's program and threw many of his colleagues completely off-kilter. Hilbert himself grumpily refused to acknowledge Gödel's result at first; eventually he accepted it and tried unsuccessfully to incorporate it into his program. Other philosophers published misguided critiques that Gödel refused to acknowledge. The English philosopher Bertrand Russell was never entirely comfortable with Gödel's result. Ludwig Wittgenstein, who completely dominated philosophical thought during the first half of the twentieth century, simply dismissed the incompleteness theorem as irrelevant. But it wasn't, and Gödel knew it.

Gödel loved Vienna, but he eventually found himself drawn to what Einstein called "a wonderful piece of Earth and . . . ceremonial backwater of tiny spindle-shanked demigods." Over a series of visits in the 1930s, he slowly began to feel comfortable at the Institute for Advanced Study, befriending Einstein, discussing with von Neumann, coming to realize just how high the intellectual caliber of émigrés ensconced in Princeton was. Following a particularly nasty incident in Vienna in which he was beaten up for looking like a Jew, he made the jump.

Einstein and Gödel hit it off immediately. As Einstein said, he would go into the office "just for the privilege of walking home with Kurt Gödel." When Gödel fell ill, Einstein came and looked after him. When Gödel applied for American citizenship and was about to be sworn in, he found what he perceived was a logical inconsistency in the American Constitution that could allow the country to descend into tyranny. Einstein stepped in and went with Gödel to prevent him from sabotaging his own citizenship ceremony.

While Gödel's obsession was mathematics, he enjoyed physics and would often spend hours discussing relativity and quantum mechanics with Einstein. Both of them found the randomness in quantum physics hard to accept, but Gödel wouldn't stop there: he thought there seemed to be a crucial flaw in Einstein's general theory of relativity.

Gödel threw himself into Einstein's field equations, and, just like Friedmann, Lemaître, and many others before him, he tried to simplify them, looking for a manageable solution that might still represent the real universe. You may remember that Einstein assumed that the universe was full of stuff—atoms, stars, galaxies, whatever might take your fancy—evenly distributed everywhere. At any moment in time, you could move around in the universe and it would look the same, completely featureless, with no center or preferred place. Friedmann and Lemaître had each in his own way followed Einstein's lead, and both of them had found simple solutions in which the geometry of the whole universe evolved with time. Gödel decided to add a small complication, small enough that he would still be able to solve the field equations but significant enough that something interesting might happen. He assumed that the whole universe was rotating around a central axis, like a merry-go-round, spinning around and around over time. The spacetime in the new universe that Gödel found, just like the universe that had been proposed by Friedmann and Lemaître, could be described in terms of time, three space coordinates, and the geometry at each point in spacetime. But there were differences. For a start, Friedmann and Lemaître's universe had the redshift effect that Slipher and Hubble had shown to be there in the real universe. Gödel's universe didn't. Quite clearly, it couldn't explain the expansion that had been measured by Slipher, Hubble, and Humason. But that wasn't the point. It was still a valid solution, a possible universe in Einstein's general theory of relativity.

However, Gödel's solution differed dramatically from all the universes that had come before in one unusual way. An observer in the Friedmann and Lemaître universe could roam around, exploring different parts of spacetime, and as time moved on, she would get older, leaving her past life behind her. There would be a clear sense of past, present, and future. This was not so in Gödel's universe. There, if an observer was moving around fast enough, she could coast along the rotating spacetime and loop back on herself. With enough accuracy, she could intercept herself when she was much younger, before she had set on her trek. In other words, in Gödel's universe, it was possible to travel back in time.

In Gödel's fantastical universe, it was possible to zoom backward and forward in time, revisit the past, correct youthful mistakes, apologize to long-departed relatives, warn yourself about future bad decisions. But it also meant that it was possible to do things that don't make sense, giving rise to some troubling paradoxes. Suppose you speed up and go back in time to meet your grandmother when she was a young girl, and through some terrible act, you kill her. You erase her existence from the face of the Earth so she can't give birth to your father or mother. You have also negated the possibility of your own existence, which means there wouldn't be a you to go back and do the dreadful deed. Yet if you lived in Gödel's universe, there was nothing to stop you from doing so, technological limitations and moral quandaries aside. Gödel's result showed that Einstein's general theory of relativity allowed for a solution in which it was possible to travel back in time and hence where paradoxes like this would be allowed, greatly at odds with our experience of the world. If Einstein's theory truly reflects nature, Gödel's absurd universe is a real physical possibility.

Gödel presented his results at a meeting to honor Einstein on his seventieth birthday, in 1949. His result was beautifully put together with a few simple statements and the final solution. But it was so outlandish that no one knew what to make of it. Chandra, who had spent the previous twenty years fending off Eddington's criticism and attacks, wrote a short note pointing out what he thought was a mistake in Gödel's derivation. But this time it was the meticulous and careful Chandra who had made a mathematical error. H. P. Robertson, an astronomer at Caltech who, along with Friedmann and Lemaître, had pioneered the expanding universe, reviewed the state of the field a year later and disparagingly dismissed Gödel's universe.

And Einstein? Einstein applied his fabled intuition, which had played such a crucial role in all his great discoveries, from special to general relativity. It was, of course, the same intuition that had made him dismiss Friedmann and Lemaître's solution and ignore Schwarzschild's. He responded to Gödel's work by saying Gödel's universe was "an important contribution to the general theory of relativity," but he reserved judgment on whether it should be "excluded on physical grounds."

Gödel's solution of Einstein's field equations seemed too bizarre to have any real bearing on the natural world. Until he died in 1978, Gödel continued looking for evidence in astronomical data that might prove that his solution had real physical significance. But in some sense, Gödel's solution exemplified the problem that so many had with general relativity: it was a mathematical theory with bizarre mathematical solutions that had no bearing on the real universe.

When the Institute for Advanced Study first tried to hire Robert Oppenheimer in 1935, just as his vibrant Berkeley school was beginning to make a name for itself, he turned it down. After a short visit, he wrote to his brother saying, "Princeton is a madhouse: its solipsistic luminaries shining in separate and helpless desolation. Einstein is completely cuckoo." He was never able to completely shake his misgivings about Einstein's later work.

In 1947, Oppenheimer finally accepted a position leading the institute. His appointment was not without opposition. Einstein and Hermann Weyl both campaigned for the Austrian physicist Wolfgang Pauli, the man who had discovered the exclusion principle, a cornerstone of quantum physics. They lobbied the faculty, stating categorically that "Oppenheimer has made no contribution to physics of such fundamental nature as Pauli's exclusion principle." But such was Oppenheimer's aura and brilliance as an organizer that he was offered the job and set about reinvigorating the atmosphere. He led with exuberance and panache. A *Time* magazine cover article in 1948 reported, "The guest list at Oppie's hotel this year will also include Historian Arnold Toynbee, Poet T. S. Eliot, Legal Philosopher Max Radin — and a literary critic, a bureaucrat and an airlines executive. There was no telling who might turn up next: maybe a psychologist, a Prime Minister, a composer or a painter." Desolate it was not.

Oppenheimer had lost interest in the general theory of relativity after his brief incursion with his students at Berkeley. He and his student Hartland Snyder had been responsible for one of the most important papers in general relativity, the discovery of collapsing spacetime. Later he had grown disenchanted with what he believed was a stale, esoteric theory, and he discouraged the young cohorts of the institute from working on it. A young member of the institute, Freeman Dyson, wrote home during Oppenheimer's reign that "the general theory of relativity is one of the least promising fields that one can think of for research at the present time." Until a new experiment could reveal more of the strange nature of space and time or someone could incorporate the general theory of relativity into quantum theory, Einstein's theory was not of much further use.

Oppenheimer was not the only leading physicist to dismiss general relativity. The rise of quantum theory had eclipsed Einstein's theory to such an extent that it had become difficult to publish papers on general relativity. The editor of the *Physical Review* was Samuel Goudsmit, a German scientist living in America who had played an instrumental role in the early years of quantum theory. Goudsmit had immigrated to America and, on becoming the editor of the *Physical Review,* had endeavored to transform it into the premier journal of physics, in direct competition with journals in Europe. Goudsmit took a dim view of general relativity. Like Oppenheimer, he felt that not much had been done or could be done with such an esoteric theory with limited applicability and testability. He threatened to put out an editorial effectively banning the publication of papers on "gravitation and fundamental theory." It was only the appeal of John Archibald Wheeler, a Princeton professor who had begun to see the charms of Einstein's theory, that held Goudsmit back from clamping down on general relativity.

Oppenheimer and Einstein eventually developed a tenuous friendship, cordial but not intimate, punctuated by acts of loyalty and affection. One time Oppenheimer surprised the old man for his birthday by having a radio mast installed in Einstein's house on Mercer Street so Einstein could hear his beloved music in the evening. Oppenheimer found in Einstein an ally who supported him through what would be his darkest days. Oppenheimer had gone through a meteoric rise during his Berkeley years and had shown spectacular stewardship during the Manhattan bomb project. He had become a firm member of the establishment as a member of the seven-man board of the Atomic Energy Commission, overseeing the development of postwar atomic projects and uses of atomic energy. He ruffled a fair number of feathers by being reluctant to endorse some of the more outlandish nuclear projects, such as a nuclear airplane that could fly continuously or the construction of the "superbomb," or H-bomb, that would have dwarfed the power and might of the atom bombs of Hiroshima and Nagasaki. In doing so, Oppenheimer made enemies. And these enemies struck back during the anti-Communist hysteria of the 1950s McCarthy era.

In a 1953 article in *Fortune* magazine, Oppenheimer was severely criticized for his "persistent campaign to reverse US Military Policy" and accused of masterminding a plot to hold back the development of the H-bomb. That year, Oppenheimer was stripped of his security clearance and deemed a threat to US security. He appealed for a hearing in 1954 and his reputation was partially cleared, but he was unsuccessful at reinstating his security clearance. As the report for the hearing clearly stated, "We find that Dr. Oppenheimer's continuing conduct and associations have reflected a serious disregard for the requirements of the security system." Oppenheimer was jettisoned from his position as a member of the Washington elite.

Einstein never understood Oppenheimer's fascination with power. Why was Oppenheimer so interested in being what amounted to a senior civil servant? As a standard-bearer for world pacifism, Einstein could not fathom why Oppenheimer, who was sympathetic to his cause, would not want to be more vocal, more public about his disapproval of the arms race. Einstein did not hold back, appearing on television to address the nation, railing against the evils of the "superbomb," which led to newspaper headlines such as "Einstein Warns World: Outlaw H-Bomb or Perish."

In his last and loneliest days, Einstein was famous yet again. Seen from afar, the situation was ironic. On one floor of the institute, Einstein would be helping to draft pacifist screeds against the proliferation of nuclear weapons. On another floor, Oppenheimer would be poring over the plans for the H-bomb. Yet Einstein could afford to be vocal. He was too famous to be touched by the anti-Communist hysteria.

So, while Oppenheimer, the key figure behind America's nuclear hegemony, was dethroned and humiliated by the security hearing, and remained cautious about not appearing to be aligned with the Communist threat, Einstein threw all caution to the wind. He publicly vilified the hearings, writing in a letter to the *New York Times:* "What ought the minority of intellectuals do against this evil? Frankly, I can only see the revolutionary way of non-cooperation in the sense of Gandhi." He proceeded to publicly advise those who were being subpoenaed by the hearings to refuse to participate by invoking the Constitution's Fifth Amendment, the right to not answer questions.

Einstein's last years were shadowed by illness. In 1948 he was diagnosed with a potentially fatal aneurysm of the abdominal aorta. The aneurysm grew slowly over the years, and Einstein prepared himself for the inevitable. When he reached his seventy-sixth birthday in 1955, Einstein realized he was too ill to travel to Bern for a conference celebrating the fiftieth anniversary of his special theory of relativity. In mid-April, his aneurysm finally burst, and after a few days in the hospital, Einstein died.

The funeral was brisk and unceremonious. A smattering of people attended his cremation, and his ashes were scattered privately. A few photographs remain of his funeral, revealing it to be a quiet, practical affair. His brain was saved for posterity in the hope that it might hold a clue to the source of his brilliance. The Bern conference went ahead, now a eulogy as well as a celebration of Einstein's work.

Oppenheimer, as head of the institute, was repeatedly asked to comment on Einstein's life and work. And he did so, praising Einstein's achievements. When pushed, he found himself unable to hide his slight disapproval of Einstein during his final years. While he had no problem in saying that "Einstein was a physicist, a natural philosopher, the greatest truly of our time," in a 1948 article for *Time* magazine on the institute, he had also been responsible for feeding the journalist a less glowing tribute: "in the close-knit fraternity of physicists, it is sadly recognized that Einstein is a landmark, not a beacon; in the quick progress of physics, he has been left some leagues behind." In an interview with *L'Express,* almost a decade after Einstein's death, Op-

Chapter 6

—————

Radio Days

BBC RADIO LISTENERS IN 1949 were duly impressed by Fred Hoyle's lectures, broadcast as a series called *The Nature of the Universe*. Here was an articulate young Cambridge don reaching out to millions of people, teaching them about the history and evolution of the universe. Like Einstein, Lemaître, and many others before him, he was bringing relativity to the masses, and the masses were enjoying it. Not yet forty years old, Hoyle could have been the new poster boy for general relativity, someone to succeed Einstein, Eddington, and Lemaître.

But Hoyle was saying that Lemaître was wrong. According to Hoyle, a universe that expanded from nothing was nonsense, and the grand old men of general relativity should have fixed the theory to get more sensible results. He claimed it was ridiculous to assume that the universe started all of a sudden. As he put it, "These theories were based on the hypothesis that all the matter in the universe was created in one Big Bang at a particular time in the remote past." He used the expression "Big Bang" disparagingly; he thought there was a much better solution: an endless universe that kept regenerating itself in a steady state of matter creation.

Hoyle was going into battle with the relativists, and with so many listeners, he was doing it from a position of strength. To the BBC's general audience, his steady-state theory sounded like the standard lore of cosmology, and the expanding universe that had come out of the successes of the 1920s seemed a renegade theory. This simply wasn't true. Hoyle and his two collaborators, Hermann Bondi and Thomas Gold, were a group of mavericks distorting the public's perception of what was really going on in theoretical physics, which deeply angered their colleagues. As one astronomer said about the response to Hoyle's lectures, there was "a feeling that he had gone far beyond the limits of decent presentations of astronomy, and a fear that his immodesty and one-sidedness had harmed the profession."

Despite Hoyle's media appeal, his steady-state theory would never be more than a cottage industry, a cult centered at Cambridge. Yet the questions that the steady-state theory raised, the young scientists that it inspired, and the new observational window onto the universe it offered would be key in the regeneration of the general theory of relativity in the decades that followed.

It is not that surprising that such a maverick as Fred Hoyle would emerge in Cambridge, the land of Arthur Eddington. Somewhat like Einstein, Eddington had also lost his way later in life and found himself obsessed with his own very esoteric theory of the universe. In the decades leading up to his death, Eddington had tried to come up with a fundamental theory that would bring everything together: gravity, relativity, electricity, magnetism, and the quantum. To an outsider, his world of numbers, symbols, and magical connections seemed more like numerology and arbitrary coincidences than the elegant mathematics at the heart of general relativity. More so than Einstein, Eddington had been shunned, spending the last few years before his death in 1944 in relative isolation. He left behind an incomplete manuscript, published posthumously in 1947 with the grand title *The Fundamental Theory*. It is an obscure book, unreadable and completely forgotten, a sad legacy from the man who had helped bring relativity to the fore. As one astronomer said at the time, "Whether or not it will survive as a great scientific work, it is certainly a notable work of art." Wolfgang

Pauli, the inventor of the exclusion principle that had been so important for understanding white dwarfs, looked at Eddington's work with disdain. To Pauli, Eddington's fundamental theory was "complete nonsense: more precisely, as romantic poetry, not as physics."

Fred Hoyle arrived in Cambridge in 1933, when Eddington was developing his theory of stars and fighting with the young Chandra over the ultimate fate of heavy white dwarfs. A round-faced, spectacled Englishman, Hoyle had first read Eddington's popular science book *Stars and Atoms* when he was only twelve years old. It was a counterpoint to what he felt was a completely inadequate education during which, as he put it, "I was allowed to drift, more or less." Yet at Cambridge he flourished, winning a number of prizes as an undergraduate and going on to complete a PhD in quantum physics. By 1939, Hoyle had been made a fellow of St. John's and had won a prestigious research fellowship. He also decided to switch fields, giving up his work on quantum physics to try his hand at astrophysics. Inspired by Eddington's *Internal Constitution of the Stars,* Hoyle decided to think about how stars burn and get their fuel. His later work would be key in understanding how nuclear processes in stars would lead to the formation of heavier elements.

When Hoyle switched fields in 1939, he was also confronted with the start of the Second World War. For the next six years, he would commit himself to the war effort, conducting radar research for the military. Just as the American atom bomb project had attracted the brightest US thinkers, the development of radio-wave technology in radar soaked up some of Britain's brightest talent during the Second World War. An array of dazzling and brilliant ideas were put to practical use for detecting airplanes, boats, and submarines. The legacy of the wartime radar effort is still with us — modern society is awash in radio waves. We use them for radio and television, for wireless networks and mobile phones, for flying aircraft and guiding missiles.

Through his work on radar, Hoyle met two young physicists, Hermann Bondi and Thomas Gold. Bondi, a Jewish Viennese émigré, had, as a sixteen-year-old, attended one of Eddington's public lectures in Vienna. He had felt compelled to move to Cambridge to study mathematics, where, completely enamored of the intellectual environment, he later wrote, "I wanted to live for the rest of my days." Coming from an enemy nation, Bondi had been interned in Canada during the initial stages of the Second World War and met Thomas Gold, another Jewish Viennese émigré who had also been enthralled by Eddington's popular books and also studied engineering at Cambridge. Once released from their internment, both Bondi and Gold worked with Hoyle on the war effort. In their spare time, they would discuss the new developments in cosmology and astrophysics, each engaging in his own way: Hoyle was bullish, Bondi was mathematical, and Gold was pragmatic.

When the war ended, all three men returned to Cambridge to take up fellowships at different colleges. Cambridge had become a harsher, emptier place after the war. Quite a few of the faculty had left, drawn by their wartime experience into pursuing careers outside academia. But real estate was at a premium, with rents driven up by the influx of workers during the war effort. Bondi and Gold ended up sharing a house just outside town. Hoyle would often spend the week in their spare room and return to his house in the countryside only on the weekends.

During the evenings, Hoyle would make the most of the extra time with Bondi and Gold, badgering them into discussing the issues that occupied his thoughts. As Gold described it, Hoyle "would continue . . . sometimes being rather repetitious, even aggravating, drumming away at particular points without any obvious purpose." One of Hoyle's obsessions was Hubble's observation of the universe's rate of expansion.

In the years since Hubble and Humason had measured the de Sitter effect, Friedmann and Lemaître's expanding universe had become firmly entrenched in the standard lore of astrophysics. While Lemaître's primeval atom was too esoteric and removed from observations to be completely adopted, it was generally felt that his model for the universe was broadly correct — the universe had been expanding from some initial time, and the details of how it began would be ironed out at a later date. It was, without a doubt, a huge success for astrophysics and the general theory of relativity.

There was, nevertheless, a baffling problem with Friedmann and

Lemaître's universe that didn't seem to go away. It had been apparent from the moment Hubble made his groundbreaking measurement. Hubble had found that the expansion rate of the universe was approximately 500 kilometers per second per megaparsec. This meant that a galaxy that was about a megaparsec distant from us (roughly 3 million light-years) would be speeding away from us at 500 kilometers per second. One that was 2 megaparsecs would be speeding away at 1,000 kilometers per second. And so on. Hubble's subsequent measurements seem to confirm this value. From this number, now known as the Hubble constant, it was possible to use Friedmann's and Lemaître's models for the evolution of the universe, wind back the clock, and figure out the exact moment in time when the universe came into being. And by doing this it was possible to work out that the universe was about a billion years old.

A billion years might seem like quite a long time, but in fact it was simply not long enough. In the 1920s, radioactive dating had determined that the Earth was over 2 *billion* years old. And work by the astronomer James Jeans seemed to pin the age of clusters of stars at hundreds to thousands of billions of years. The ages of star clusters were later revised down, but there was no doubt about it: it seemed as if the universe was *younger* than the stuff it contained. That simply couldn't be true, but there seemed to be no way around the paradox. Willem de Sitter summed up the situation in 1932 by saying, "I am afraid all we can do is to accept the paradox and try to accommodate ourselves to it." The situation hadn't improved by the time Hoyle, Bondi, and Gold became interested in the expanding universe.

When the Cambridge trio started thinking about cosmology, the age paradox seemed like a glaring failure of Friedmann's and Lemaître's models. But what really troubled Hoyle, Bondi, and Gold was something much deeper and much more conceptual. In winding back the clock of Friedmann's or Lemaître's model the beginning of the universe corresponds to a moment when the whole of space is infinitely concentrated at a single point. In other words, space, time, and matter came into being at that one, initial instant. To Hoyle and his friends this was anathema. As Hoyle would put it, "It was an irrational process that cannot be described in scientific terms." What laws of physics could be used to describe the creation of something out of nothing? It was inconceivable, and for Hoyle it was "a distinctly unsatisfactory notion, since it puts the basic assumption out of sight where it can never be challenged by a direct appeal to observation." Their dismissiveness echoed Eddington's appalled assessment of Lemaître's primordial egg.

It was a movie, *Dead of Night,* that led Hoyle and his colleagues to take a fresh look at the universe. Made in 1945, *Dead of Night* is a horror movie with a circular structure, the ending neatly matching the beginning. With no real beginning and end it is a claustrophobic vision of an endless universe. And it intrigued Hoyle, Bondi, and Gold. What if the universe was in fact like that? There would be no initial time, no primordial egg.

Bondi and Gold viewed the problem of the initial time — or as Hoyle would later call it, the "Big Bang"— from an almost abstract, aesthetic point of view. Over the centuries, descriptions of the universe had moved away from having special, preferred positions in space. Friedmann and Lemaître, like Einstein before them, postulated that the universe was completely featureless, with no center or preferred place from which things evolved or were observed. There was true democracy among all points in space. So why not promote this principle, the cosmological principle, to something far more complete and all-encompassing? Why not assume that all points in space *and* moments in time were the same? There would be no beginning, just an eternal universe that would remain in a steady state for all time.

Hoyle set about figuring out the details of such a proposal. In Friedmann and Lemaître's universe, energy would be diluted with the expansion and slowly decrease with time. If the universe was to be truly in a steady state, the energy would have to be replenished somehow to keep the universe chugging along. And so Hoyle decided to fix Einstein's field equations, much as Einstein had tried to do when he was constructing his now-defunct static universe. Hoyle postulated the existence of what he called the creation field, or C-field as it became known, that would create energy over time. Hoyle's steady-state universe would be sustained by this mysterious source of energy, which had never before been seen. In Hoyle's universe one of the sacrosanct laws of physics — conservation of energy — went down the drain.

Hoyle argued it wasn't that big a deal, for all that was needed was "about one atom every century in a volume equal to the Empire State Building." Almost nothing.

Two papers, one by Hoyle and another by Bondi and Gold, came out in 1948 in the *Monthly Notices* of the RAS. The reception was mixed. One of the fathers of quantum physics, Werner Heisenberg, who had stopped by in Cambridge when Hoyle presented his C-field paper, thought it was the most interesting idea to come out of his visit. E. A. Milne, an Oxford professor of mathematics, rejected it outright, stating, "I do not believe the hypothesis of the continual creation of matter is necessary, nor do I consider that it is on the same footing as the assumption that the universe as a whole was created at a particular epoch." Max Born, who had supervised Robert Oppenheimer in Göttingen, simply couldn't stomach the changes Hoyle was proposing, "for if there is any law which has withstood all changes and revolutions in physics, it is the law of conservation of energy." And the great man himself, Albert Einstein, paid little attention to Hoyle's model, claiming it was simply a piece of "romantic speculation." What seemed to the trio of astronomers like a simple, obvious solution to such a fundamental problem in cosmology was being dismissed as outlandish and unnecessary. Hoyle was frustrated by what he perceived as the unreasonableness of his colleagues. As he put it, he was quite "worn out with explaining points of physics, mathematics, fact and logic to obtuse minds."

And then, an opportunity to promote his model that would far surpass the impact of any paper or seminar series landed in Hoyle's lap. The BBC was planning a series of radio lectures by the Cambridge historian Herbert Butterfield. Butterfield pulled out at the last minute and the young Fred Hoyle, who had some experience of broadcasting, was invited to take Butterfield's place and record a series of programs on the universe and cosmology, five in total. In them Hoyle could expound on the problems of cosmology, the young universe with old galaxies, and how Friedmann and Lemaître's universe created more problems than it solved. And he could describe the virtues of his steady-state universe. Hoyle could bypass all the conventional methods and present his ideas to the whole country as a fait accompli. Everyone would know about his theory.

Hoyle's BBC lectures were incredibly successful and Hoyle became a well-known figure, one of the first media dons. The public warmed to his description of the universe, and it took hold in the popular imagination. But by taking such a public stage to promote his own model above the much more well-established and accepted expanding universe discovered by Friedmann and Lemaître, Hoyle rankled his colleagues, and the concept of a steady-state universe suffered a backlash as a result. While Hoyle had succeeded in placing the steady-state universe on a public stage, resistance among his colleagues became more firmly entrenched. As Hoyle later recalled, "I found it difficult to get my papers published during the first two or three years of the 1950s."

Nevertheless, the steady-state universe took hold as a viable alternative to the expanding universe of Friedmann and Lemaître that had won over Einstein. The great discoveries of the 1920s in cosmology and general relativity were under assault. But in the next few years, a completely new window on the universe would open up and cast all these models in a different light.

"I do not think it unreasonable to say that [Martin] Ryle's motivation in developing a program for counting radio sources . . . was to exact revenge," recalled Hoyle of his former colleague. It was an uncharitable thing to say, but there was definitely an element of truth in it. For Martin Ryle was a volatile, irascible character, competitive and suspicious. Even within Cambridge, Ryle would isolate himself from the rest of the faculty, going to work near the radio telescopes based at what used to be the Lord's Bridge railway station, "in a shed in the fields," as one of his colleagues recalls. He would have a distinguished career — he would become the Astronomer Royal in 1972 and win the Nobel Prize in 1974 — yet throughout, Ryle behaved as if he were constantly under threat, enforcing a bunker mentality in his group.

Martin Ryle had also come out of the radar generation. The son of a Cambridge professor, Ryle graduated from Oxford in 1939 with a first-class degree. Like Bondi, Gold, and Hoyle, Martin Ryle worked on radar during the war, coming up with tricks for jamming the German radar systems and subverting German rocket guidance systems.

After the war Ryle went to Cambridge, where he set about applying his skills to developing, and at some point dominating, the new field of radio astronomy. He was not alone, for when Bernard Lovell, who had also spent the war enmeshed in the development of radar, moved to Manchester, he set about building one of the largest steerable radio telescopes in the world at the Jodrell Bank Observatory. In Australia, Joseph Pawsey spent his war years developing radar for the Royal Australian Navy before setting up his own radio astronomy group in Sydney.

The first steps in radio astronomy had been taken a few years before, when Karl Jansky, an engineer working for Bell Telephone Laboratories in New Jersey in the early 1930s, realized that the universe was hissing at him. Jansky had been asked to find the source of the annoying static that was making conversations over the radio and even broadcast radio programs sometimes impossible to hear. Jansky just wanted to fix the radios — he had little interest in the mysteries of outer space.

Radio waves behave just like light waves, but their wavelengths are a billion times longer than those of visible light. The light we can actually see, which makes up the bulk of the sun's rays, has a wavelength that is less than a millionth of a meter. Radio waves have gigantic wavelengths, ranging from a millimeter all the way up to hundreds of meters. Jansky had found that the Milky Way was emitting an extraordinary amount of radio waves, day in and day out. Even though the sun was much brighter in the sky than the whole Milky Way put together, it didn't emit as many radio waves. In an article "Electrical Disturbances Apparently of Extraterrestrial Origin," published in 1933, Jansky systematically took apart all possible sources of static and showed a map of where the radio waves were coming from. His methods revealed a different way of looking at the cosmos. Instead of using giant telescopes with lenses on mountaintops, this kind of observation could be done with chicken wire, steel, and dishes out in the open plains. Rather than looking at the faint light of distant objects, astronomers could pick up the radio waves coming from outer space.

Jansky's discovery was mostly ignored. When he proposed that Bell Labs build a new, improved antenna, he was refused. They weren't in the business of astronomy. And so Jansky moved on to other things. But his work wasn't completely forgotten. An idiosyncratic radio engineer and amateur astronomer from Wheaton, Illinois, by the name of Grote Reber read about Jansky's discovery in *Popular Astronomy* and set about building a bigger and better antenna in his backyard in Wheaton. Reber's antenna had a 9-meter dish, with metal scaffolding that extended out in front to capture the reflected waves. It was the first proper radio telescope, much like the ones we see today. With it, Reber set out to make a finer map of the radio emissions of the Milky Way and build a detailed map of the radio sky. He submitted his work to the *Astrophysical Journal* where Chandra, who was the editor at the time, was intrigued by Reber's results and bemused by his persistence — he accepted the paper for publication. And so in 1940, Reber's "Cosmic Static" was published with his very own maps.

Reber's new radio maps of the Milky Way were interesting, helping to map out in detail where all the mysterious waves were actually emanating from. But Reber's measurements also revealed something else: a few isolated points on the maps were beaming copious amounts of radio waves. While Reber was able to place each of the points near a constellation — Cygnus, Cassiopeia, and Taurus — they did not correspond to objects emanating visible light. Reber had discovered a new type of astronomical object that became known as a radio source or radio star.

"Cosmic Static" opened up a new window on the universe. Unfolding before a new generation was perfectly uncharted territory, and Martin Ryle was ready to explore. Along with Lovell's and Pawsey's groups, from the late 1940s onward, Ryle and his group at Cambridge began mapping the cosmos. Deploying the techniques that he had learned while working on radar, Ryle designed a new generation of radio telescopes that would transform Cambridge into one of the premier centers for radio astronomy. But it also would bring him up against Hoyle and his collaborators.

Martin Ryle was more of a radio-ham amateur and an electrical engineer than a cosmologist, so it was surprising that he would get caught up in a fight with "theoreticians," as he would disparagingly call Hoyle and his colleagues. But he had walked right into it. He had first tried to find more bright radio sources, like those Reber had observed, and pinpoint their locations, but unfortunately he made the wrong call. It seemed clear to him that all these objects were firmly embedded in the Milky Way. In a clearly argued paper in 1950 he made the case that the majority of radio sources should lie within our galaxy. There could be a few odd outliers, but on the whole they must be close by. What he said made sense and was entirely reasonable.

Ryle presented his results at a meeting of the Royal Astronomical Society in 1951. In the audience were his Cambridge colleagues Gold and Hoyle, who stood up and casually conjectured that the radio sources might actually be extragalactic. Ryle, who had carefully thought through his arguments, was annoyed and dismissed Gold and Hoyle, saying, "I think the theoreticians have misunderstood the experimental data."

It was a clash of cultures pitting the highbrow theoretical astronomers, versed in mathematics and physics with elegant yet odd theories that explained the whole universe, against the tinkerers, the radio operators who built kits and played with electronics. Ryle couldn't stand the perceived condescension of his colleagues. He felt he understood the data in a way that these people who worked solely with pencil and paper couldn't. Unfortunately for Ryle, Gold and Hoyle were eventually proved right as more and more radio sources came to be associated with objects outside the Milky Way. They were indeed extragalactic, and Ryle had to accept that the theorists *did* in fact understand the data.

But Ryle didn't accept defeat quietly. Given that these radio sources lay outside the galaxy, they could be used to say something about the universe. So Ryle turned to amassing more observations and using his data to go after Hoyle and Gold's baby, the steady-state theory. He did so by counting the number of radio sources as a function of their brightness and trying to relate this number to the underlying properties of the universe. The farther away a radio source is, the dimmer it will be, so the dimness of a source can be seen as an indicator of its distance. The universe is a big place and there is a lot of space out there, so one would expect to see more dim, distant sources than bright, close ones. It turns out that the ratio of the number of dim sources to bright ones is a good way of figuring out what type of universe we might live in. When we look at distant sources, their light has taken time to reach us, so we are looking at the universe when it was younger. If we live in Hoyle, Gold, and Bondi's steady-state universe, the density of sources remains constant over time, so the total number of sources within a certain volume should be directly proportional to that volume. In an evolving universe like the one Friedmann and Lemaître proposed, the universe was denser in the past than it is now, so there should more distant, dimmer sources than close, bright ones. By counting the number of dim sources relative to the bright sources, it should be possible to determine whether our universe adheres to the Big Bang or the steady-state model.

Ryle compiled a list of almost two thousand sources in what was called the 2C Catalogue (*C* stands for Cambridge). It built on a much smaller list of fifty sources (known as the 1C Catalogue) and seemed, to Ryle's satisfaction, to have far too many dim sources compared with bright sources for it to be consistent with the steady-state theory. Ryle saw this as the killer blow for Hoyle's theory and immediately set about advertising his results. In a prestigious lecture he was invited to give at Oxford in May 1955, he came out with a bold indictment of his rivals: "If we accept the conclusion that most of the radio stars are external to the galaxy, and this conclusion seems hard to avoid, then there seems to be no way in which the observations can be explained in terms of a steady state theory." Ryle had seemingly demolished Hoyle and Gold's model.

After Ryle's lecture at Oxford, Hoyle and his collaborators were on the defensive. Hoyle took the data seriously, but Gold was suspicious of the results, advising Hoyle, "Don't trust them, there might be lots of errors in this and it can't be taken seriously." Gold was right. This time Ryle was thwarted by his own cohort, the same tinkerers who were building radio astronomy into a bona fide science. Two young Australian radio astronomers, Bernard Mills and Bruce Slee from Sydney, reanalyzed the 2C data and found a completely different result from Ryle's. Instead of trying to come up with a catalogue of thousands of sources to rival Ryle's, they opted to focus on a small subset of the whole survey, about three hundred sources, and measured them in exquisite detail. This small catalogue was picked so it overlapped with Ryle's catalogue and could be used to actually check Ryle's measurements.

Mills and Slee's published results completely destroyed the credibility of Ryle's survey. In their paper, they said that their "catalogue is compared in detail with a recent Cambridge catalogue . . . it is found that they are almost completely discordant." Mills and Slee went on to suggest that "the Cambridge catalogue is affected by the low resolution of their radio interferometer." Ryle's results were simply not good enough — Mills and Slee were working with a better telescope that was more precise, and their results could not exclude the steady-state model as a possible model of the universe. A radio astronomer from the rival group in the UK named Jodrell Bank chimed in, saying, "Radio astronomers must make considerable progress before they can offer the cosmologists anything of value." It seemed that the radio astronomers couldn't agree on their data, let alone use it to test cosmological models, so it was deemed best to ignore that data for now. Hoyle and his collaborators had a field day.

Ryle retreated to Cambridge to work on the next generation of his source catalogue. Singed by the debacle of his questionable results, Ryle and his team spent the next three years building a new catalogue, unimaginatively called the 3C Catalogue. The new results would decisively shoot down the nonsense that Hoyle and his team were peddling, or so Ryle thought. In 1958, when the 3C Catalogue was finally revealed to the world, Martin Ryle finally felt he had his *pièce de résistance:* a collection of radio sources that everyone agreed with. Yet it still wasn't good enough. Bondi was skeptical and pointed out that Ryle had a tendency to claim that his measurements were better than they actually were; Ryle often claimed to have ruled out the steady-state model when really he had just reached the limits of what could be said with his data. Whenever anyone went back and reanalyzed Ryle's data and found that the errors were larger than previously claimed, the steady-state model was ruled back in the game. Indeed, as Bondi said publicly, "this has happened more than once in the last ten years."

In February 1961, Ryle presented his analysis of what was now the 4C Catalogue at the Royal Astronomical Society meeting. He argued that the results were simply incompatible with those of the steady-state model — there were far too few bright sources relative to the dim ones. The observations, he said, "appear to provide conclusive evidence against the steady state theory." The newspapers picked up on Ryle's announcement and came out with headlines claiming that "the Bible was right" about the existence of an initial moment of creation. As other teams in Australia and the United States reproduced Ryle's results, it seemed that Ryle had finally sorted it out.

Hoyle and his collaborators were worried but not convinced. As Bondi told the *New York Times,* shortly after Ryle announced his analysis, "I certainly don't consider this the death to continuous creation," adding, "A similar statement has been made by Professor Ryle in 1955 but the observations on which it was based were later found to be incorrect." There was something irrational about Ryle's personal quest to kill the steady-state theory, even though the data was improving year by year. To Hoyle, Bondi, and Gold, radio hadn't killed the steady-state theory, at least not yet.

The fight between Hoyle and Ryle, centered in Cambridge as it was, may seem like an unnecessary distraction from the inexorable progress of general relativity and cosmology. Few people outside the United Kingdom had any interest in Hoyle's model. To many, the debate seemed fickle, almost unscientific, driven by personalities and vendettas. Visitors to Cambridge would comment on the poisonous atmosphere between Ryle and Hoyle's group.

But their rivalry resulted in significant scientific progress. Fred Hoyle would go on to be lauded as one of the great astrophysicists of the second half of the twentieth century. With William Fowler and Geoffrey and Margaret Burbidge from the United States, he would end up developing a brilliant theory for the origin of the elements in stars. Some might point to his maverick nature and his insistence on supporting the steady-state model to explain why he was not included as one of the awardees of the 1983 Nobel Prize in Physics. In 1973 he left Cambridge to live in the Lake District and write novels.

Hermann Bondi would end up creating a vibrant general relativity group in King's College London, and Thomas Gold would end up setting up the world's largest radio telescope in Arecibo in Puerto Rico. Martin Ryle's group developed a reputation for secrecy and paranoia, yet they were behind some of the great discoveries in radio astronomy of the following two decades. Ryle won the Nobel Prize in 1974. The rise of radio astronomy and the elusive nature of radio sources would play a crucial role in the advancement of general relativity, which was about to enter a new phase.

Chapter 7

Wheelerisms

JOHN ARCHIBALD WHEELER personally discovered relativity by way of nuclear physics and quantum theory. In the spring of 1952, Wheeler found himself wondering what happened at the end of the lives of stars made of neutrons, the building blocks of nuclear physics that Wheeler had spent his life until then studying. He was puzzled by Robert Oppenheimer's prediction that the endpoint of the gravitational collapse of such a star could be a singularity, a point of infinite density and curvature at the star's center. To Wheeler, these singularities didn't sound right. They couldn't be truly physical, and there must be some way to avoid them. To understand this bizarre prediction, Wheeler would have to learn general relativity. He figured the best way to do that would be to teach it to the students at Princeton. And so, in 1952, in the home of Einstein, Gödel, and Oppenheimer, John Archibald Wheeler taught the first course on general relativity in the Princeton department of physics. Until then it had been considered an abstract subject more suitable for a mathematics department. It was a momentous departure, one Wheeler would recall years later as "my first step into a territory that would grip my imagination and command my research attention for the rest of my life."

Wheeler was, as one of his students summed him up, a "radical con-servative." He definitely looked conservative, always dressed impec-cably, in a dark suit and tie, hair perfectly groomed, his shoes shined, the perfect image of a traditional, somewhat conventional gentleman. Devoted to his students and collaborators, he was courteous and polite with an old-fashioned sense of propriety. Yet he would say the most outlandish things, often spouting out cryptic phrases about cosmic enigmas that made him sound more like a New Age guru or an en-lightened hippie.

As a scientist, Wheeler saw himself as both a dreamer and a "doer." His interests ranged from the esoteric to the practical. He was just as fascinated by explosives and mechanical devices as he was by the magical new rules of atomic theory. At university, Wheeler studied engineering and discovered the full glory of mathematics. One of his mathematics teachers had given him advice on how to tackle prob-lems; as Wheeler recalled, he "liked to tell us in class while teaching us new mathematical tricks that an Irishman gets over an obstacle by going around it." This advice affected Wheeler's approach to solving problems throughout his life. He would fearlessly jump into problems, learning whatever he needed, when he needed it. In 1932 he was only twenty-one and already had a PhD in quantum physics.

John Wheeler came of age as a quantum physicist when all the great discoveries of Schrödinger and Heisenberg were bearing fruit. As a young faculty member at Princeton, he worked with the Danish phys-icist Niels Bohr on the quantum properties of nuclei and how they interact. Wheeler and Bohr's work on nuclear fission was published on exactly the same day as Oppenheimer and Snyder's work on gravi-tational collapse and played an important role in the lead-up to the Manhattan Project.

Wheeler's conservatism came out in his passionate belief in the American way of life, its institutions, and its defense. He joined the atomic bomb project immediately after Pearl Harbor, working on the giant reactors that were needed to create plutonium for the bombs. His brother died in combat in 1944, and for the rest of his life Wheeler felt that he hadn't done enough in pushing more actively for the develop-ment of the bomb much earlier. As he would later tell his colleagues, if the bomb had been developed earlier, it could have been used sooner, in Germany. The loss of life would have been tremendous but not, in his view, as terrible as in the last year of the war. His patriotism sometimes put him at odds with his colleagues. In the early 1950s, he was invited to work with Edward Teller on the Matterhorn Project, an attempt by the United States to develop the H-bomb, a thermonuclear weapon that would work with nuclear fusion. He did so even though many of his colleagues, including Robert Oppenheimer, were firmly against it. Wheeler was one of the few physicists to stand back from supporting Oppenheimer when he faced charges of breaching national security.

Though conservative politically, he couldn't resist being a maverick, or radical, in science, following outlandish ideas that went against the physics establishment at the time. Among Wheeler's students at Princeton was Richard Feynman, a brilliant young man from New York who would come to be the poster boy of postwar quantum physics. Under Wheeler's supervision, Feynman would come up with a completely revolutionary way of explaining and calculating how particles and forces played against each other in the arena of spacetime. It was Wheeler who encouraged Feynman to think differently and be bold.

Wheeler was the perfect person to pick up the pieces of general relativity. He was both practical and visionary. He was conservative and respectful of the physics and astrophysics that had led up to the theory but at the same time keen on trying different, new, untested approaches. And above all, he was an inspiring mentor who would train and support a new generation of physicists who would breathe new life into the general theory of relativity.

Once Wheeler had taught himself general relativity, he embraced it. It was too elegant, and the experimental facts, meager as they might be, too compelling, for the theory not to be true. But that didn't mean he was opposed to testing its limits. He believed that "by pushing a theory to its extremes, we also find out where the cracks in its structure might be hiding," and so he set out to discover just how strange general relativity could be. In the process he often assigned pithy, simple one-liners to his outlandish ideas, popularly known as Wheelerisms.

One idea, developed with his talented student Charles Misner, was to incorporate electric charges into general relativity without actually having any charge whatsoever. "Charge without charge" was the Wheelerism he came up with to describe the concept. The thought experiment used a series of mathematical tricks to poke holes in two distant pieces of spacetime and connect them with a tube of spacetime they called a wormhole. Through these tunnellike wormholes they could thread electrical field lines. Field lines emerging from one end of the wormhole would make it act as if it had a positive charge, attracting negative charges toward it. The field lines entering at the other end would make it appear to have a negative charge. The wormhole would act like a pair of positive and negative charges very far apart, but, in fact, no charged particles were involved. It was an ingenious idea, easy to visualize but fiendishly difficult to work with in practice.

"Mass without mass" was another Wheelerism. Einstein's theory explains how objects with mass interact, but Wheeler wanted to find a way to derive Einstein's results without involving any mass at all. In Einstein's theory, light can bend space just like mass, so Wheeler proposed that if he could compress a bunch of light rays in a way that warped space and time enough, they would look just like a mass. The bundle of light, or geon as he called it, would have weight and attract other geons. The light rays would have to be wrapped up into a doughnut-shaped coil and could be easily taken apart, but they would have the effects of mass without actual mass. With another student, Kip Thorne, Wheeler set out to determine if these objects could exist in nature without immediately becoming unstable.

And then there was, of course, the problem of marrying the quantum to general relativity. It was too good a problem, and too extreme, for Wheeler to resist an attempt at solving it. Once again, he made things up. Wheeler postulated that if you were to observe spacetime on the smallest scales, you would start to see strange effects emerge. While spacetime might look smooth at large scales, mildly warped by the presence of massive objects (including Wheeler's geons and wormholes), at smaller scales you would see a roughness that you hadn't realized was there. With a really powerful microscope you might find that spacetime is a turbulent mess, all jumbled up. In fact, quantum uncertainty should make spacetime look like a roiling foam on the smallest scales. It is only because we see the world with blurry vision that we are unable to observe its fundamentally rough nature.

Yet while Wheeler embraced the outlandish and was happy to propose bold scenarios, he felt deep unease about the singularities lurking at the core of the work of Schwarzschild, Oppenheimer, and Snyder on collapsing massive stars, which had first sparked his interest in general relativity. To Wheeler, the strange singularities *must* be a strange mathematical artifact that wouldn't arise in nature. As Wheeler recalled, "For many years this idea of collapse to what we now call a black hole went against my grain. I just didn't like it."

So he set about trying to *fix* the problem by inventing new physical processes that would come into play when collapse pushes matter into the obscenely high densities at a star's core. This was new territory for him, for although Wheeler had become one of the world's experts in nuclear physics, the physics that described neutrons at the center of gravitational collapse was a completely different matter. He needed to figure out what would happen when neutrons were packed far more densely than in Landau's or Oppenheimer's neutron stars or in any bomb he might have come up with during his work for the American military. It was a form of guesswork and imagination at which he excelled. But despite his creativity, and like Landau and Oppenheimer before him, Wheeler and his group also found that there was a maximum mass above which not even their detailed, speculative proposals for the end state of matter would compete with gravity. Whatever they did, it simply wasn't possible to avoid the formation of a singularity at the end of gravitational collapse. Yet Wheeler simply couldn't stomach the singularity and refused to give in.

As Wheeler grew more and more fascinated with general relativity and his quest to rid it of singularities, he cajoled his students and postdocs to join him on his journey. Like their mentor, they were seduced by the power of the theory, intrigued by what could be done. Year after year, Wheeler's group came up with new ideas, some outlandish, some sensible, but all of them captivating. Wheeler's influence on general relativity extended well beyond Princeton. One of his greatest contributions was his quiet support for Bryce DeWitt at the University of North Carolina at Chapel Hill.

There was something formidable about Bryce DeWitt. He had a towering, stern presence, like an Old Testament prophet, and when he walked into a room, backs straightened. He had no time for sloppiness — things had to be done properly, so when ideas finally made their way to paper and publication, they were set in stone.

DeWitt was also a traveler, a "space traveller," as he liked to call himself. As a young man, he was a pilot in the Second World War, and following his graduate studies at Harvard he hopped around the globe, working at Princeton, Zurich, and the Tata Institute in Bombay, the latter of which a colleague later described as "a sojourn [that] did not make good professional sense, but . . . suited his roving spirit."

DeWitt settled in California with his wife, Cécile DeWitt-Morette, a French mathematician he had met at Princeton, and found work at the Lawrence Livermore Laboratory developing computer simulations for modeling nuclear artillery shells. But the family needed more money to buy a house, so one evening, DeWitt decided to enter an essay competition that offered a first prize of $1,000. The essay would change everything — not just for the DeWitts, but for general relativity.

The Gravity Research Foundation competition was the brainchild of Roger Babson, a businessman who was passionate about gravity. He made his fortune playing the stock market by applying his own version of Newton's laws of physics: "What goes up will come down. . . . The stock market will fall by its own weight." It wasn't rocket science, but Babson was a man with an obsession. His older sister had drowned when he was a young child, and he blamed gravity. In his version of the tragedy, "she was unable to fight gravity, which came up and seized her like a dragon." Throughout his life Babson invested in gravity in one way or another, collecting Newton memorabilia, promoting outlandish ideas, and, most significantly, creating the Gravity Research Foundation.

Babson originally envisioned the Gravity Research Foundation as the sponsor of an annual essay-writing competition. Candidates would submit essays of no more than two thousand words suggesting ways to harness gravity and achieve Babson's ultimate goal: antigravity. The foundation would lead to the development of antigravity devices: contraptions that could insulate, absorb, or even reflect gravity. The atom was being harnessed, and Babson thought it was time for gravity to be brought under control as well. He intended for his essay competition to bring out the best in postwar physics.

The initial response to Babson's challenge was lackluster. From 1949 until 1953, essays trickled in with a few muted suggestions. The topics were scattered, and the contestants were a mixture of academics, graduate students, and amateurs racking their brains to come up with something that might fit Babson's requirements. But the topic was too outlandish, bringing cranks out of the woodwork instead of inspiring real science.

Babson's challenge certainly wasn't respectable — no right-minded physicist actually believed that it would be possible to build an antigravity machine — but it did echo a wider growing interest in the potential of gravity. After the Second World War the US economy was booming and optimism was infecting everyday life. It was the beginning of the atomic era, the birth of a new technological age. With money to invest, organizations and business people placed substantial bets on gravity as the next big thing after nuclear energy. There was something truly appealing and revolutionary about a goal that, in essence, came directly out of a science fiction novel. For it was, if anything, an attempt to do what H. G. Wells had written into his 1901 novel *The First Men in the Moon,* to discover the magical substance "cavorite" that could reverse gravity and take people to the moon.

Throughout the mid-1950s there were routine references in the broadsheets to a new form of space travel that could beat gravity. Articles with headlines like "Space Ship Marvel Seen If Gravity Is Outwitted," "New Air-Dream Planes Flying Outside Gravity," and "Future Planes May Defy Gravity and Air Lift in Space Travel" marveled at a future with "gravity propulsion systems." The popular press imagined airplanes or spacecraft that would use gravity instead of jet engines to propel them forward. A *New York Herald Tribune* article with the headline "Conquest of Gravity Aim of Top Scientist in the U.S." described how aviation companies like Convair, Bell Aircraft, and Lear, Inc., were looking into gravity, which just might "eventually be controlled like light and radio waves."

The Glenn L. Martin Company (which later became known as Lockheed Martin) set up the Research Institute for Advanced Studies. The institute would explore new ideas in theoretical physics with a special emphasis on unpicking gravity and pursuing gravity propulsion, hiring physicists and relativists to help them pursue their futuristic goal. The US Air Force made a more sober and less outrageous investment in the Aeronautical Research Laboratory, based at the Wright-Patterson Air Force Base in Dayton, Ohio. The ARL also housed a group of bona fide relativists, but they conducted fundamental research in gravity and unified theories. There was no mention of antigravity in their remit, and for a while the research group at ARL was a proper general relativity research center that rivaled the few other groups scattered around the world. The air force also pumped money into other groups that were conducting general relativity research. Few scientists took antigravity efforts seriously, and researchers avoided making any ludicrous predictions, but they happily accepted the money thrown at them to focus on esoteric ideas about the foundations of reality.

In the midst of this euphoria, Bryce DeWitt's approach to Babson's contest was certainly a strange way to win an essay competition; he attacked the sponsors. In the essay he submitted to the Gravity Research Foundation in 1953, DeWitt brazenly dismissed Babson's ambitious aims to develop "grossly practical things such as gravity reflectors or insulators or magic alloys which can change gravity into heat." He invoked Einstein's theory of spacetime to explain why "any frontal attack on the problem of harnessing the power of gravity along the above lines is a waste of time. . . . One may safely pronounce all gravity-power schemes impossible." DeWitt slammed the cranks, and he won.

DeWitt's essay was definitely different from those of the previous contestants. It was proper science, firmly stepping away from speculation and talking about the real scientific issues that needed to be faced in gravity research. It was a hard task, and, as he said, "gravitation has received relatively little attention during the last three decades." It was "peculiarly difficult," involved "recondite mathematics," and the "fundamental equations are almost hopeless of solution." Indeed, "the phenomenon of gravitation is poorly understood even by the best of minds."

Far from insulted, Roger Babson was intrigued by his competition's first real contender. Here was someone serious, a proper scientist who could make his competition reputable. And indeed, DeWitt's essay added legitimacy to Babson's competition, for in the years that followed the caliber of the contestants went up dramatically. In fact, over the following decades many of the physicists who would play a crucial role in the resurgence of general relativity would end up winning prizes from the Gravity Research Foundation. Moreover, the essays became almost exclusively about gravity, and antigravity was forgotten. DeWitt would later say that winning that competition was "the quickest $1000 I ever earned," but, having taken part in the competition, DeWitt was to benefit much, much more than he had imagined.

Roger Babson had a friend, Agnew Bahnson, who was also fascinated by gravity. Bahnson had made his fortune selling industrial air-conditioning units. Like Babson, he wanted to fund research in gravity. He just wasn't sure how. Babson showed Bahnson DeWitt's winning essay. Here was the man to help him set up something serious, a proper, respectable institute where thinkers would be allowed to follow their interests. As Bahnson wrote in one of his inaugural brochures for the newly created Institute of Field Physics, or IOFP for short: "In the minds of the public the subject of gravity is often associated with fantastic possibilities. From the standpoint of the institute no specific practical results of the studies can be foreseen at this time." There would be no antigravity machines, no gravitational propulsion. Bahnson could satisfy his personal fantasies about gravity another way, by writing science fiction, and leave real gravity to the scientists.

Bahnson turned to John Wheeler for advice on how to proceed with his institute. Wheeler had earned a formidable reputation in Washington for his work on nuclear weapons and more generally as a senior physicist who was willing support the government in all matters related to defense. He had followed DeWitt's career at a distance and quietly supported the idea that Bryce and Cécile should be invited to be the first researchers at the new institute, based in Chapel Hill, North Carolina.

The institute may have started as a vanity project, but with Wheeler's backing and the DeWitts as the first hires, it was taken seriously by scientists from all across the country, with letters of support from many of the *éminences grises* applauding a place where pure research could be undertaken, unfettered by the demands of industry, the army, or the new atomic age. At the core of the new institute would be gravity.

The DeWitts' January 1957 meeting, titled "The Role of Gravitation in Physics," was intended to inaugurate the new institute. It also inaugurated a new era. The group of attendees was younger and less well known, but they included some of the new leaders in general relativity. They all converged on Chapel Hill for a few days to take Einstein's theory apart. Agnew Bahnson and the US Air Force funded it, and the air force even flew some of the participants over to the newly founded Institute of Field Physics.

Not only relativists made the trip to Chapel Hill. John Wheeler's ex-student Richard Feynman, who had completely overhauled quantum physics and proposed a new way of quantizing nature, decided to attend. A man from the quantum world, he was intrigued by what was going on in general relativity. Feynman later recalled arriving at the airport in Chapel Hill without knowing where to go. Once in a cab, he realized that the driver hadn't heard of the meeting — why would he have? Feynman turned to the driver and said, "The main meeting began yesterday, so there were a whole lot of guys going to the meeting who must have come through here yesterday. Let me describe them to you: They would have their heads kind of in the air, and they would be talking to each other, not paying attention to where they were going, saying things to each other, like 'gee-mu-nu. gee-mu-nu.'" Gee-mu-nu (written $g_{\mu\nu}$) is the mathematical symbol for the metric that encodes the geometry of spacetime. The driver knew where to go.

It was apparent to everyone at the meeting that something had to be done to pull the general theory of relativity out of the backwater where it had been languishing for the past three decades. To Richard Feynman, it was obvious why general relativity had been neglected: "There exists . . . one serious difficulty, and that is the lack of experiments. Furthermore, we are not going to get any experiments, so we have to take the viewpoint of how to deal with the problems where no experiments are available." Without experiments, the field could not progress, but Feynman insisted they had to press on. General relativity was difficult but not *that* difficult and, as he put it, "the best viewpoint is to pretend that there are experiments and calculate. In this field we are not pushed by experiments but pulled by imagination."

Feynman echoed the general feeling in the meeting at Chapel Hill, which was full of a new generation of relativists who were about to graduate or had just graduated with new ideas, ready for a fight. As the meeting unfolded, outlandish ideas competed against sober pronouncements by the older pundits. The daily sessions were riven with debate and arguments. When Thomas Gold presented an update on his theory of the steady-state universe, DeWitt chipped away at its key premise — Hoyle's creation field — questioning the mechanism by which energy conservation would be violated. When someone played up the need for a theory that unified gravity and electromagnetism along the lines that Einstein had spent decades trying to construct, Feynman was unforgiving. Why should electromagnetism be the only force that needs to be unified with gravity? What about everything else, all the other forces in nature? DeWitt and Wheeler's obsession, how general relativity could be combined with quantum mechanics, was aired and discussed in its various forms and guises. And could spacetime ripple with gravitational waves like the surface of a lake, just like electromagnetic waves in Maxwell's theory? The participants fought it out in the lively discussion sessions.

John Wheeler turned up with his grand plan to revolutionize physics through relativity and with his cohort of students and postdocs presented their new ideas. They pushed relativity even further than before, to the point where it seemed like a joke. On the menu was "electromagnetism without electromagnetism" and "charge without charge," as well as "spin without spin" and "elementary particles without elementary particles." Throughout the meeting, the Wheeler clan took center stage, throwing ideas into the crowd to be thoughtfully considered or batted away. John Wheeler was in his element.

At an even more basic level, the relativists at Chapel Hill asked themselves if it was even possible to make realistic predictions with Einstein's theory. If a theory is going to have any cachet, it must be predictive. So, for example, electromagnetism is supremely successful at predicting just about everything that pertains to light, electricity, and magnetism. But, while Schwarzschild, Friedmann, Lemaître, and Oppenheimer had all been able to make predictions, they had restricted themselves to highly simplified, idealized systems. And it wasn't clear how to go beyond those simplifications. Indeed, the participants of the Chapel Hill conference asked themselves, Was it even possible to *generally* solve the Einstein field equations properly and make real bona fide predictions about how spacetime evolves? It seemed that the hideously entangled nature of general relativity makes just choosing the initial conditions, let alone the evolution, almost impossible. Attempting to solve the equations on a computer was an even more daunting task.

The meeting was an exciting forum for relativity's new adherents, bursting with creativity and invigorated by John Wheeler's inventiveness and Feynman's imagination. But the theory of spacetime was still stuck. All the mathematical ingenuity, the proposals for unification, the debates about gravitational waves, and Wheeler's wormholes, geons, and spacetime foam were useless if they couldn't be pinned to the real world.

It had been almost forty years since Eddington's eclipse measurement, the first big test of Einstein's theory. Almost thirty years had passed since Hubble's measurement of the expansion of the universe. At the Chapel Hill meeting there were no new measurements, nothing to further confirm or even unsettle Einstein's theory. One of Wheeler's Princeton colleagues, Robert Dicke, summed up the situation in a talk on "The Experimental Basis of Einstein's Theory" when he said, "Relativity seems almost to be a purely mathematical formalism, bearing little relation to phenomena observed in the laboratory." The answer, as it turned out, was to be found not in the laboratory but in the stars.

In 1963, the Dutch astronomer Maarten Schmidt had the run of a telescope named after George Ellery Hale, the patron of the Palomar observatories. On his mind was one of the sources in the 3C Catalogue of radio astronomers Martin Ryle and Bernard Lovell. While Wheeler and his crew were reenergizing general relativity, radio astronomers were taking a closer look at the radio sources in their surveys. As with any other stargazers, their goal was to figure out what the radio sources actually were. To do so, they needed to find more of them. And they needed to look at them more carefully to figure out what was actually emitting the radio waves.

Over more than a decade, deploying the ingenuity that had helped them develop radar, Ryle and Lovell increased the precision of their measurements by orders of magnitude, enabling them to pinpoint the radio sources in the sky exactly enough for astronomers to point their ordinary telescopes at them and figure out what they were. Ryle's 3C Catalogue of radio sources included hundreds of sources with precise locations.

Lovell's group looked at Cygnus A, one of the radio sources that Grote Reber had identified above the cosmic static emanating from the galaxy and dubbed 3C405 in Ryle's catalogue. Cygnus A turned out to be a strange object, consisting of two lobelike blobs of radio waves, each one almost rectangular in shape. They were gigantic structures, each one hundreds of light-years across, and seemed to be powered by something lying between them. When astronomers pointed their telescopes at another source called 3C48, instead of finding the intricate structure they had found around Cygnus A, they saw a simple bright spot dominated by light in the blue end of the spectrum. It looked like a star, simple and featureless. But when they tried to measure spectra to figure out what 3C48 was made of, the forest of spectral lines that read off their instruments couldn't be matched to any of the stars they knew, nor could they identify any of the elements that it was made of. There were many other objects they couldn't identify. Cosmic radio sources were plentiful and different, and no one had a clue what or how far away they were.

Maarten Schmidt focused on a source that had the nondescript name of 3C273. It looked like a star, but the spectral lines were unlike any set he had seen before. Looking closely at his measurements, he found something quite remarkable: the spectral lines of the radio source matched those of hydrogen exactly if they were dramatically redshifted by almost 16 percent. Line by line he could match the two spectra. But to be redshifted by that amount, 3C273 was either hurtling away from us at speeds close to the speed of light or was so far away that the expansion of the universe was dramatically redshifting the spectra. Schmidt was stunned. That evening he told his wife, "Something terrible happened at the office today."

It was a momentous discovery. Schmidt had found that these objects littered throughout the cosmos were billions of light-years away, and for such distant objects to be seen so easily in radio surveys and by large optical telescopes, they had to be belting out an enormous amount of energy. In fact, 3C273 and 3C48 were producing as much light as one hundred galaxies put together. They were like supergalaxies, much more powerful than anything that had been seen before.

These sources also had to be very, very small, only a fraction of the size of any other galaxy. The same was true of other sources in the 3C Catalogue — some were ten or even a hundred times smaller than ordinary galaxies. And when monitored closely, these sources seemed to be less than a few trillion kilometers across, "mere peanuts by cosmological standards," as *Time* magazine wrote at the time. Copious amounts of energy were being produced at colossal distances from a very small region of space.

Something that inexplicable and bizarre was too tempting for Fred Hoyle. While continuing his battles defending the steady-state model of the universe, Hoyle had developed a formidable reputation as an expert on the structure of stars. With William ("Willy") Fowler and Geoffrey and Margaret Burbidge, he had come up with a detailed explanation of how the elements in nature could all be synthesized in nuclear reactions in stars.

Fowler and Hoyle proposed that the radio stars were indeed stars, but not like any other stars. These stars would be *superstars,* with masses of a million or a hundred million suns like ours, so immense that they could produce tremendous amounts of energy during their lifetimes. And their lifetimes were short, for they burned up their energy so quickly that they would rapidly collapse in a brief, violent death. With their superstars, Hoyle and Fowler pushed the rules for understanding stars developed by Eddington well into the realm of the general theory of relativity. Einstein's theory beckoned.

In the oppressive heat of the summer of 1963, a small group of relativists gathered in Dallas, Texas. They sat around the pool drinking martinis and discussing the strange, heavy objects that Maarten Schmidt had unlocked. They were an international bunch in Dallas for, as one of them put it, "American scientists outside of geophysics and geology would rarely deign to settle there. To most the region seemed to be as magnetic as Paraguay." But Texas was to become an unlikely center for relativity, a shift driven mainly by the efforts of a hard-talking, gregarious Viennese Jew named Alfred Schild.

Schild had an itinerant childhood and youth, a product of the turmoil of the 1930s and 1940s. He was born in Turkey and lived in England as a child. Like Bondi and Gold, he was interned in Canada, where he studied physics under Leopold Infeld, one of Einstein's disciples, and wrote a thesis on cosmology. He had been at the meeting in Chapel Hill in 1957, taking part in the excitement of general relativity's next phase, and that year he was recruited to take up a professorship at the University of Texas at Austin.

Texas was a backwater when Alfred Schild arrived in Austin, but it was phenomenally rich from the oil income that was flowing through the local economy. Schild was able to cajole the university to put the oil money to good use, letting him set up his own Center for Relativity. With the air force keen to tap into the potentially magical powers of gravity, there was no shortage of money. And while the mathematicians at Austin looked down on Schild's work, the physicists were willing to take him in.

Schild went looking for talent, and he definitely had a knack for finding it. The group of young relativists he assembled from Germany, England, and New Zealand transformed Austin, Texas, into an obligatory stopping point for any relativist worth his salt. Schild didn't stop in Austin. In Dallas, the newly created Southwest Center for Advanced Studies was looking for young faculty to boost the "science starved south," so Schild stepped in. Schild told them to invest in relativity, and so they did, hiring the center's very own international group to build up the ranks of Texan relativity.

That July afternoon, the Texan relativists lounging by the pool cooked up a scheme that would bring the world to Texas to discuss relativity. It wouldn't just be another Chapel Hill, small and freewheeling. This time they would bring in a whole new crowd, the astronomers, and try to rope them into thinking about Einstein's theory by hosting a meeting focusing on radio stars, the "quasi stellar radio sources." With Schmidt's measurements of the previous March, it was clear that these strange objects were too massive and too distant to be treated using the old Newtonian laws of gravity. These were the big things that Chandra and Oppenheimer had alerted everyone to, the stars that would be too big to withstand the pull of gravity, and where general relativity could play such a dramatic role. In the invitation letter they sent out, the organizers proposed that "energies which lead to the formation of radio sources could be supplied through the gravitational collapse of a superstar." The relativists called their meeting the Texas Symposium on Relativistic Astrophysics. It was to be held in December of 1963 in Dallas.

The first Texas Symposium on Relativistic Astrophysics was almost canceled. President John F. Kennedy had just been assassinated in Dallas, and conference goers were simply too scared to come to Dallas and run the risk of being shot. The Dallas relativists asked the mayor to reach out to potential attendees individually and assure them of the city's safety. It worked. Over three hundred people turned up in Dallas to hear the latest about radio stars and what could be made of them. Among the crowd was Robert Oppenheimer, who had discouraged work on general relativity at the institute in Princeton. He was intrigued by these new radio stars, for they were, as he described them, "incredibly beautiful ... spectacular events of unprecedented grandeur." He commented on how the meeting resembled those in quantum physics almost two decades before "when all one had was confusion and lots of data." For him, it was an exciting time.

The meeting went on for three days, with astronomers and relativists alike debating the import of the strange "quasi stellar radio sources" in Ryle's 3C Catalogue. One of the meeting's attendees starting calling them "quasars," which was quicker and easier to pronounce. For the relativists, these quasars seemed so massive and so concentrated that Schwarzschild's weird solution, and Oppenheimer and Snyder's calculation, had to be taken into account if any sense was to be made of the data. The astronomers and astrophysicists found the quasars so bizarre and mysterious that they started paying attention to what the relativists were saying. Maybe, just maybe, general relativity had to be brought into the picture to make any sense of these new discoveries.

At Dallas, more than ten years after he had started working on general relativity, John Wheeler was present and ready to say his piece. The big unanswered question on his mind was what he called "the issue of the final state." He wanted to find out what happens at the endpoint of gravitational collapse. He still found it impossible to believe Oppenheimer and Snyder's prediction that singularities formed, and he was convinced general relativity would play an integral role in explaining why they wouldn't. Despite his prejudice, he felt duty-bound to explain all the possibilities and enlist his audience in his pursuit of the final state. Before his talk, Wheeler picked up a piece of chalk and meticulously filled a blackboard with his elaborate pictures and equations illustrating what he had been thinking about for almost a decade. On the board were plots showing how he thought a star would collapse under its own weight and how general relativity predicted the star's inexorable movement toward its final fate. Scattered around were equations, bits of Einstein's field equations, summaries of quantum physics, a hodgepodge of brilliance that helped him lay out his results of the past ten years. More than anything, Wheeler's talk was an apologia of general relativity arguing that it should be taken seriously by any right-minded astrophysicist.

For many of the astronomers the results were too fanciful, and one of the attendees recalled "utter disbelief" on the face of "a distinguished participant." Yet others marveled that the universe had finally caught up with Wheeler. It seemed the general theory of relativity that he had been thinking about for so long now actually had relevance and might be of use to understand the new radio observations.

In a description of the meeting, *Life* magazine said, "The scientists, having stretched their imagination to a point that once would have embarrassed science fiction writers, were hardly less mystified than they were before they began their talks . . . so fantastic is the nature of radio sources that no bets were ruled out." During the after-dinner speech, Thomas Gold summed up the extraordinary turn of events that they were witnessing at the symposium: "Here we have a case that allowed one to suggest that the relativists with their sophisticated work were not only magnificent cultural ornaments but might actually be useful to science! Everyone is pleased: the relativists who feel they are . . . suddenly experts in a field they hardly knew existed; the astrophysicists for having enlarged . . . their empire by the annexation of another subject — general relativity." He ended on a cautious note, saying, "Let us all hope that it is right. What a shame it would be if we had to go and dismiss all the relativists again."

With his incredible vision and persistence, John Wheeler had overseen the resurrection of Einstein's moribund theory. By devoting his fearsome intellect and creativity to training a new generation of brilliant young relativists, and supporting the new centers that were scattered throughout the country, he had nurtured a new and vibrant community that could think deeply about gravity. Finally, the data had been obliging, and with astronomers, physicists, and mathematicians ready to tackle the big questions, the Texas Symposium heralded a new era. General relativity was back.

Chapter 8

Singularities

WHILE MOST OF the audience listened to John Wheeler's presentation at the 1963 Texas Symposium with incomprehension, one young mathematician watched enthralled as Wheeler lectured in front of his carefully prepared blackboard of equations and plots. "Wheeler's talk made a real impression on me," Roger Penrose recalls. And even though Wheeler stubbornly refused to accept the existence of singularities, he was, in Penrose's mind, asking the right question: Could these singularities be an essential ingredient of general relativity? Wheeler's talk at the Texas Symposium heralded the start of a decade that would be dubbed the "Golden Age of General Relativity" (by one of Wheeler's own students, Kip Thorne), and Roger Penrose would be one of the brilliant thinkers to see it through.

Penrose has spent his life playing with spacetime: cutting it up, gluing it back together, pushing it to its limits. He sees things differently, possessing a mathematician's gaze enhanced by a more visceral understanding of space and time. His drawings, known as Penrose diagrams, unwrap spacetime and reveal its oddest properties. They visualize what happens to light as it zooms past the Schwarzschild surface, how light behaves as you follow it back to the Big Bang, and even how space and time can be stretched to look like the frothy surface of the sea.

Penrose was still an undergraduate, studying mathematics in London, when he first felt the pull of general relativity. He taught himself the basics using a book by Erwin Schrödinger aptly called *Space-Time Structure*. But what really set him thinking about the details were Fred Hoyle's lectures proselytizing about his steady-state theory. There was something fascinating but also odd about the universe that Hoyle was describing — it didn't fit with Penrose's understanding of relativity. He decided to pay a visit to his brother Oliver, also a mathematician, who was studying for a PhD in Cambridge. He thought Oliver could help him understand this strange theory that so appealed to him.

Cambridge in the 1950s, despite the staid atmosphere of centuries-old cloisters and the stifling rituals of the colleges and university, was becoming an exciting place. Paul Dirac, an English physicist who had played a crucial role in showing that the quantum theories of Heisenberg and Schrödinger were one and the same, gave brilliant, exquisitely crafted lectures on quantum mechanics. Hermann Bondi lectured on general relativity and cosmology and, with Fred Hoyle, actively promoted their steady-state universe. And then there was Dennis Sciama.

Penrose and his brother met at the Kingswood restaurant in Cambridge to discuss Fred Hoyle's radio lectures. Penrose simply couldn't understand Hoyle's claim that in the steady-state model, galaxies would speed up and away so quickly that at some point they would disappear over a cosmic horizon. He recalls thinking that something else ought to happen, something he could show with his diagrams. Oliver pointed over to another table and said, "Well, you can ask Dennis. He knows all about it." He walked Roger Penrose over to Dennis Sciama and introduced them. They hit it off immediately.

Sciama was only four years older than Penrose but was already embroiled in Einstein's theory with a passion that he would pass along to a string of students and collaborators over almost fifty years. He had done a stint at the Institute for Advanced Study in the year before Einstein died. In one of his few conversations with Einstein, Sciama had boldly, and somewhat rashly, declared that he was there to "support the 'old Einstein' against the new." Einstein had laughed at his impudence. Sciama had studied with Paul Dirac, to the extent that such a thing was possible, and had become seduced by the work of Hoyle, Bondi, and Gold. Yet while he was a staunch believer in the steady-state universe, he paid attention to what the radio astronomers were finding. The results coming out of Ryle's group down the road intrigued him. He could see how they might sink Hoyle's model.

That evening in the Kingswood, Penrose explained to Sciama why galaxies wouldn't disappear from sight. They would get dimmer and, from a distance, would appear to freeze in time, just as Oppenheimer and Snyder had shown would happen with an imploding star as its surface passes through the Schwarzschild horizon. Sciama saw the spark in Penrose's eyes and loved his fresh approach to looking at spacetime. They would be friends for the next fifty years.

Penrose eventually moved to Cambridge to pursue a PhD in mathematics, but he remained beguiled by the mathematical oddities he'd found in the geometry of spacetime. He desperately wanted to understand them better. When he finished his PhD, he took the plunge and decided to work on general relativity. He spent the next few years roaming the world, working with Wheeler in Princeton, Hermann Bondi in London, and Peter Bergmann in Syracuse. He finally joined Schild's Austin, Texas, group in the autumn of 1963.

Texas was the hot spot for general relativity, and researchers there were flush with funding. "We didn't really ask where the money was coming from or why anyone thought it was worthwhile to spend all that money on relativity," Penrose says. "I always felt there must be some mistake." One of Penrose's colleagues was a young New Zealander named Roy Kerr. Kerr had spent long days in the Texas heat and humidity grappling with Einstein's field equations, trying to find more complex, more realistic solutions. He had come up with an elegant set of equations that corresponded to a simple geometry for spacetime. Kerr's solution could be seen as a more general form of Schwarzschild's geometry. While Schwarzschild described a spacetime that was perfectly symmetric around a point, the point where the infamous singularity would lie, Kerr's solution was symmetric around a line that cut through the whole of spacetime. It was as if he had set Schwarzschild's solution spinning on an axis, twisting and tugging spacetime around it. If he wanted to retrieve Schwarzschild's original solution, all he had to do was stop his solution from spinning.

Penrose immediately took to Kerr's result. He spent hours discussing the discovery with his new colleagues at Austin, rephrasing the new spacetime in his own way. Like Sciama, Schild was taken by Penrose's way of seeing things. Penrose's mathematical insight and diagrams shed a completely new light on Kerr's solution. Kerr submitted his remarkably simple and powerful result to the *Physical Review Letters,* the American journal that only a few years before had considered banning the publication of anything related to relativity. It was instantly accepted and published in September 1963, just a few months before the Texas Symposium was to take place in Dallas. There he could present his result to the astrophysicists.

Afraid that Kerr's presentation might be too dry and mathematical, Schild tried to convince Penrose to present the new solution instead of Kerr. Penrose would have none of it; it was Kerr's baby. Schild's concerns were not entirely unfounded. When Kerr went to the podium to make his presentation, half of the participants left the hall. Kerr was young and unknown, a relativist among a gang of astrophysicists who had better things to do at that moment. Kerr spoke to the remaining, desultory crowd, and, as Penrose recalls, "They didn't pay much attention to him." Very few people understood the point of Kerr's result, the first big step in making Schwarzschild's solution more general, more real, and more useful to astrophysicists. Kerr wrote a short note for the conference proceedings, but the person charged with summing up the main results of the symposium left him out entirely. It was still too much general relativity for the astrophysicists to accept.

There wasn't a single Soviet physicist at the first Texas Symposium. Much of the precious intellectual power of Soviet physics had been taken up with the Soviet nuclear project, leaving little time or attention for general relativity. However, just as a new generation of relativists emerged from the Manhattan Project in the United States and radar in the United Kingdom, many of the Soviet nuclear scientists would eventually lead a revival of general relativity in the Soviet Union in the 1960s.

The Soviet nuclear project was late getting started. During the Second World War, precious resources had been drained from the Soviet machine on the Soviet German front, which prevented Joseph Stalin from putting his men to work on the bomb. Starting in 1939, following John Wheeler and Niels Bohr's paper that discussed the copious release of energy from the nuclear fission of heavy elements, scientific papers on nuclear fission in the West seemed to have dried up. To the Soviets, it was as if Western research into nuclear fission had ground to a halt. In 1942, when a Soviet physicist, Georgii Flerov, wrote to Stalin and alerted him to this strange state of affairs, Stalin became suspicious. He guessed that the Americans were working on a bomb, and he realized he had to get in the game. Once the war ended, Stalin plundered his own scientific elite to set up a bomb project. The team included Lev Landau and Yakov Zel'dovich.

Lev Landau had suffered under the wave of persecutions during the great terror of the late 1930s. His stint in prison had left him a deeply bitter man, profoundly disillusioned by the regime, yet at its mercy. Landau had already become legendary, with a raft of discoveries to his name spanning from quantum mechanics to astrophysics. He had created a school of physics and a following of brilliant disciples who would be tried to the limit of their intellectual abilities just to be allowed to work with him. In fact, to be accepted as one of Landau's protégés, aspirants had to pass a series of eleven punishing exams, known as "Landau's Theoretical Minimum," set and overseen by Landau himself, a process that could take up to two years. Only a few made it through the barrier and were able to work with the great man himself.

Yakov Zel'dovich, a Belorussian Jew just a few years younger than Landau, had been a precocious student. He became a lab assistant at seventeen, gained a doctorate at twenty-four, and rapidly became one of the Soviet authorities on combustion and ignition. It was inevitable that he would be roped into developing the bomb, and he did so with flair. From 1945 until 1963, Zel'dovich took part in the construction of the first Soviet atomic bomb, dubbed "Joe-1" by the Americans when they detected its explosion in August of 1949, and then worked on its successor, the "superbomb." The Soviet Union had caught up with the Americans and become a nuclear power.

While Zel'dovich was passionate about the nuclear project, Landau, still smarting from his ordeal in the Lubyanka and nursing a profound hatred for Stalin, had been coerced into taking part. And while Zel'dovich greatly admired Landau, Landau was less charitable toward his colleague and the nuclear project as a whole. When Zel'dovich attempted to enlarge the Soviet nuclear bomb project, Landau called him "that bitch." When Stalin died, he said to a colleague, "That's it. He's gone. I'm no longer afraid of him, and I won't work on [nuclear weapons] anymore." Nevertheless, for their contribution to the Soviet bomb project, both men were awarded the Stalin Prize and the Hero of Socialist Labour medal a number of times. Landau went on to win the Nobel Prize in 1962.

In the mid-1960s, Zel'dovich's star was still rising, but Landau was incapacitated, laid low by a car crash that left him a shell of the man he once was, unable to do physics. Landau's protégés carried on in his stead; they were the first Soviets to go after singularities in spacetime. The two young men, Isaak Khalatnikov and Evgeny Lifshitz, who had both undergone the rigors of an education with Landau, were well prepared to tackle the intricacies of Einstein's theory to look at what happens when matter collapses under its own gravity.

Oppenheimer and Snyder had built their solution around a simple approximation, a perfectly symmetric sphere of stuff collapsing inward. The perfect symmetry had initially bothered people like Wheeler, who saw it as too much of an idealization. The surface of the Earth is covered with irregularities: huge mountains and deep oceans and valleys. What if a collapsing star was similarly uneven? Could the irregularities and imperfections distort the collapse so much that parts of the surface would fall in far more quickly than others, rebound, and make their way out again? If that was so, singularities might never form.

The Russians addressed this question by loosening the symmetries Oppenheimer and Snyder had enforced. In Khalatnikov and Lifshitz's calculation, spacetime could twist and churn in each direction in a different way. Imagine looking face-on at the seething mass of stuff, a massive star, for example, as it implodes, collapsing inward toward its center. In general, you would expect it to appear lopsided. The top and the bottom bits of the blob might collapse more quickly than the sides, so quickly that they might bounce right back out before the sides of the blob had time to collapse. Instead of everything falling inward, inexorably forming the singularity, there would always be some part moving outward, holding spacetime up. Only if the collapse was set up just so, perfectly symmetric around the center, would everything fall in at *exactly* the same time, allowing the singularity to form. Khalatnikov and Lifshitz's paper, published in the Soviet journal *Soviet Physics,* came to the striking conclusion that in realistic situations singularities *never* formed. Schwarzschild's and Kerr's solutions were abstractions that should never form in nature. Einstein and Eddington, it appeared, had been right all along.

Soviet scientists were occasionally allowed to attend meetings in the West. The Third Meeting on General Relativity and Cosmology, the successor of the Chapel Hill meeting, was held in London in 1965, with over two hundred relativists in attendance. When Khalatnikov presented his results there, all those relativists paid close attention. While it was evident that Einstein's theory had taken off in the Soviet Union, it was difficult for Western scientists to tell exactly what was going on. Translations of the main Soviet journal, *Soviet Physics,* were always delayed.

Penrose sat quietly and listened to Khalatnikov's presentation. He knew they were wrong but thought it would be "undiplomatic" to speak out. "You couldn't really prove anything doing it the way they did it," he says. "There were simply too many assumptions. They couldn't rule out singularities like that." In fact, Penrose could prove that singularities *always* formed, contrary to Khalatnikov's claim. Penrose's results were completely general because he had used his own new methods of looking at spacetime.

Since his first encounter with Sciama at the Kingswood restaurant in Cambridge almost ten years before, Penrose had developed his diagrams into a set of rules for how to think of light, or anything for that matter, propagating through spacetime. He could take an arbitrary spacetime, and from looking at some of its most basic properties and what kind of stuff it contained, he could get a definitive sense of what would happen to it, whether it would collapse to a point or explode out to infinity. When he applied his rules to the problem of gravitational collapse, what Wheeler called "the issue of the final state," the outcome was inevitable: singularities. Penrose wrote up his paper, "Gravitational Collapse and Spacetime Singularities," and submitted it to *Physical Review Letters.* As he summarized in his paper, "Deviations from spherical symmetry cannot prevent spacetime singularities from arising." Almost half a century later, it is still a masterpiece of concision, clarity, and rigor: a perfect paper in just under three pages, with a brief explanation of the problem, the mathematical toolkit and the proof in a small paragraph, all illustrated with one of Penrose's signature diagrams.

When Khalatnikov gave his presentation, Penrose had already submitted his paper. It was about to be accepted and would be published in December of that year, but his techniques were unfamiliar to most of the relativists in the audience, especially the Russians. When Charles Misner, one of John Wheeler's students, stood up and challenged Khalatnikov with Penrose's result, it was a lost battle. Suspicious of Penrose's result, the Russians refused to accept that there might be an error in their own approach. "I hid in the corner," Penrose recalls. "It was too embarrassing."

But Penrose was right. What became known as his singularity theorem had far-reaching consequences. It meant that if general relativity was correct, the Schwarzschild and Kerr solutions, those strange spacetimes with singularities at their centers, had to exist in the universe. They weren't merely mathematical constructs. Einstein and Eddington were wrong. Four years later, Khalatnikhov and Lifshitz admitted defeat. In 1969 they looked at their calculations again, this time with one of their students, Vladimir Belinski. To their dismay, they found a mistake. While in 1961 they had thought that the collapse that leads to the formation of a singularity was too special and unnatural to occur in the real world, with Belinski they found quite the opposite. In their own way they confirmed Penrose's theorem: singularities always formed. They humbly published their results in the West, publicly acknowledging their mistake.

Penrose had proved the inevitability of singularities in gravitational collapse and answered Wheeler's question about the final state. Deeper confirmation would soon follow.

Martin Ryle may have failed in his first attempts to dismantle Cambridge's steady-state orthodoxy through his initial radio source measurements, but his data was improving. In 1961, when he released the 4C Catalogue of radio sources, most of the radio astronomers agreed that many of the problems with the previous data had been fixed. But the end of the steady state would begin with the theory's own adherents.

Dennis Sciama was a strong advocate of Hoyle's steady-state theory. He was also fascinated by quasars and assigned one of his students, Martin Rees, the task of looking at Ryle's new measurements in different ways. Rees took a simpler and much cleaner approach than Ryle's technique of plotting the number of quasars as a function of flux. Instead, Rees took a subset of thirty-five quasars with measured redshifts and divided it up into three slices. One slice exhibited low redshift, corresponding to quasars close to Earth in time and distance. The second slice contained quasars with medium redshifts, and the final slice was made up of objects with high redshifts viewed in the distant past.

Rees's idea was simple but remarkably clever. In the steady-state model, in which the universe does not evolve over time, each slice should have approximately the same number of quasars. Instead, Rees found almost no quasars in the most recent slice. Almost all of them were in the farthest slice. In other words, the number of quasars seemed to have changed with time — there were more in the past — and so the universe couldn't be in a steady state. The plot told it all — the steady-state universe didn't work. "It was really that plot that converted Dennis," Rees recalls. From then on Sciama believed in Lemaître's theory, or the Big Bang as Hoyle had called it in his lectures, and whatever that entailed.

The final nail in the coffin of the steady-state theory came from across the pond in New Jersey. Arno Penzias and Robert Wilson had been working on an antenna at Holmdel, one of the telecommunications sites belonging to Bell Labs. They wanted to retrofit the antenna, a huge horn that captured radio waves, and use it to measure the galaxy. To accurately map out the structure of the Milky Way, they first needed to determine the precision of their instrument. So they used the antenna to stare at nothingness and check how well they could see it.

But what they saw wasn't nothing. Penzias and Wilson were definitely seeing or, to be more precise, *hearing* something: a low, soft hiss streaming out of empty space. No matter how they adjusted their instrument, they couldn't get rid of it. These two men had inadvertently stumbled upon a relic from the early universe, a fossil of the Big Bang.

In the late 1940s a Russian physicist working in the United States, George Gamow, predicted the existence of a very cold bath of light permeating the universe. He started from the Abbé Lemaître's idea that the universe started in a hot, dense soup from which all the elements eventually emerged. The argument goes as follows: Imagine a universe in its simplest state, just full of hydrogen atoms. The hydrogen atom is the elementary building block of chemistry, a proton and an electron held together by electromagnetic force. If you bombard a hydrogen atom with enough energy, you can rip the electron away from its nucleus, leaving a lone proton floating in space.

Now imagine a gas of hydrogen atoms pushed together in a hot bath. They will collide, move around, and be bombarded by energetic photons, beams of light whizzing around. And the hotter they are, the more likely it is that the electrons will rip away from the protons. If the environment is very hot, very few hydrogen atoms will remain intact. Instead of a gas of hydrogen, the universe will be full of free protons and electrons. Early in the life of the universe, when the universe's temperature was greater than a few thousand degrees, you would find very few atoms and mostly free protons and electrons. As time passes and the universe cools, electrons stick to nuclei, leaving mostly hydrogen and helium atoms, an almost insignificant smattering of heavier elements, and a faint, almost invisible background of light. This is what Arno Penzias and Robert Wilson saw — clear evidence for a hot, dense state at early times. It was as close as one could get to proving the existence of a Big Bang, as Hoyle had disparagingly called it, and it would be another of Dennis Sciama's students, Stephen Hawking, who would take that final step.

There was something of Einstein in the young Hawking, and indeed his childhood friends would often call him that. He hadn't shone at school, and if anything he had been relaxed, playful, and naughty, a slight, untidy boy who delighted in entertaining his colleagues. Hawking had become increasingly interested in science and, on applying to Oxford, had aced the entrance exam and interview. He found Oxford ridiculously easy and had done well enough to impress his tutors and lecturers. It was at Cambridge as a PhD student, under Sciama's tutelage, that Hawking would be steered toward the cosmos and, finding his scientific voice, would spell out one significant consequence of Penzias and Wilson's discovery.

Stephen Hawking was a year older than Martin Rees and became fascinated by the mathematics of general relativity. Early on during his PhD studies, he had been diagnosed with Lou Gehrig's disease and given just a couple of years to live. The initial news had been profoundly demoralizing, yet two years into his PhD he was still alive and well. His continued health galvanized him to focus on his work and try to understand what actually happened at the beginning of the universe's expansion — at the Big Bang itself. Could it be that the singularities were inevitable at the beginning of time as well as in Wheeler's final state?

As he raced against the possible onset of his illness, Hawking was able to show that, indeed, an expanding universe under normal conditions should have inevitably started off with a singularity. Over the years, he proved, with a South African physicist and fellow talented Sciama student named George Ellis, that a universe with relic radiation like that found by Penzias and Wilson must have started in a singular state. Finally, with Roger Penrose, he constructed a complete set of theorems that covered almost any possible model of an expanding universe that could be cooked up at the time. Singularities were inevitable, or so Penrose and Hawking's math seemed to say, both in the future as well as in the past.

At the first Texas Symposium, there had been speculation that the distant, copious sources of radio waves in Ryle's catalogue might somehow be related to the general relativistic collapse of supermassive stars.

Chandra had once pointed out that superheavy white dwarfs would be unstable and might implode, and Oppenheimer and Snyder had shown that if stars were even heavier, the next stage in the inexorable collapse would be via neutron stars. While there was pretty convincing evidence for white dwarfs, there was no sign of neutron stars. That changed in 1965, when Jocelyn Bell arrived in Cambridge to start her PhD in Martin Ryle's group.

Bell didn't work with Ryle himself but with one of his more junior colleagues, Antony Hewish. Hewish had her build a radio telescope out of an assortment of wooden posts and chicken wire that she could use to pinpoint and study the position of quasars at 81.5 megahertz. As she puts it, her "first couple of years involved a lot of very heavy work in the field, or in a very cold shed." But the job had its perks: "When I left I could swing a sledge hammer." By 1967, Bell was taking data on a chart recorder, analyzing over 30 meters of chart paper a day, looking for the telltale signals of quasars. About 120 meters of paper would cover the whole sky.

There was something odd in the recording she was making. For each 120 meters of paper, there was a quarter-inch spike of data that Bell couldn't understand. She couldn't figure out what the signal was or where it was coming from. It was undoubtedly there, a set of chirps in a very specific direction of the sky. "We had begun nicknaming it 'little green men,'" Bell recalls. "I went home feeling very fed up." The team decided to go ahead and publish their mysterious finding.

In February 1968, a paper appeared in *Nature* titled "Observation of a Rapidly Pulsating Radio Source." In it, Bell, Hewish, and their co-authors announced their discovery, declaring, "Unusual signals from pulsating radio sources have been recorded at the Mullard Radio Astronomy Observatory," and then went on to make a bold claim: "The radiation seems to come from local objects within the galaxy, and may be associated with oscillations of white dwarf or neutron stars." They speculated that the spikes in the chart paper were the oscillations or pulsations in these dense, compact radio sources.

The press took to the discovery, interviewing Hewish about its importance. But, as Bell recalls, "journalists were asking relevant questions like was I taller than or not quite as tall as Princess Margaret."

She says, "They'd turn to me and ask me what my vital statistics were or about how many boyfriends I had . . . that was all women were for." The *Sun* headlined the news piece with "The Girl Who Spotted the Little Green Men." It was the *Daily Telegraph* that came up with a name for the outlandish objects; a journalist suggested calling the objects "pulsars," short for "pulsating radio stars."

Yet again, radio astronomy had delivered in spades, and yet again, it was by chance. The discovery was momentous, and in 1974 the Nobel Prize was awarded to Bell's supervisors, Tony Hewish and Martin Ryle. Bell was left out entirely, in what is seen by many as one of the greatest injustices in the history of the prize. Almost twenty years later, Bell attended the prize ceremony as the guest of another astronomer, Joseph Taylor Jr., when he won the Nobel Prize in 1993. "I did get to go in the end," she recalls without any bitterness.

Pulsars were the first tangible evidence for neutron stars. They don't actually pulsate — they rotate, which causes them to emit a periodic signal. But they were the fabled missing link in gravitational collapse, posited by Landau, studied by Oppenheimer, and explored in meticulous detail by Wheeler and his disciples. They were the final step before the formation of Penrose's inevitable singularities.

When Yakov Zel'dovich switched fields, he did so fearlessly. One of his students recalls Zel'dovich's advice: "It is difficult, but interesting to master ten percent of . . . any field. . . . The path from ten to ninety percent is pure pleasure and genuine creativity. . . . To go through the next nine percent is infinitely difficult, and far from everyone's ability. . . . The last percent is hopeless," from which Zel'dovich concluded, "It is more reasonable to switch to a new problem before it is too late."

Like Wheeler, Zel'dovich turned from nuclear research to relativity in his forties, and he went on to set up one of the most focused research groups in the world. The papers that Zel'dovich wrote with his students were almost impressionistic, often with quirky openings such as "The Godfather of psychoanalysis Professor Sigmund Freud taught us that the behavior of adults depends on their early childhood experiences. In the same spirit, the problem is to derive the . . . present . . . structure of the universe . . . from . . . its early behavior." They read like condensed essays, with a smattering of equations, just enough to flesh out his point. When translated into English, they could be difficult to decipher. But over time they were appreciated for what they were: veritable gems of relativistic astrophysics.

When Zel'dovich switched fields, he went looking for frozen stars, as Schwarzschild and Kerr's collapsing stars were called in the East. These frozen stars are invisible, emit no light, and have no surface that can reflect or shine. Yet Zel'dovich couldn't accept that these strange objects would be hidden from view, for they were dramatic, distorting space and time about them. In fact, as he began to discuss with his students, they should exert an inexorable pull on anything that gets near them. And so, he surmised, by looking at the effect of the frozen stars on other things, it just might be possible to see them, not directly but indirectly. For example, if the sun got too close to a frozen star, it would be forced to orbit around it, much like the moon around the Earth. The frozen star would be invisible, so the sun would look as if it were dancing around on its own, wobbling about in a strange orbit with no center. Look for wobbling stars, Zel'dovich and his team proposed: stars that appear to be on their own but behave like half of a binary system.

But, Zel'dovich conjectured, frozen stars shouldn't just nudge their partners around; they should positively rip them apart. He made a very simple assumption: as stuff falls into the gravitational field of a frozen star, it should approach the speed of light, condensing and heating up in the process. As the material mixes and collides, heating up as it falls onto the frozen star in a process that has been dubbed accretion, it radiates energy. The accretion near the Schwarzschild horizon is so efficient it can emit up to 10 percent of its rest mass energy, an astounding amount of energy that makes it the most efficient energy-generating process in the universe. And so, in a short paper published in *Doklady Akademii Nauk* in 1964, Zel'dovich went on to speculate that the production of energy around a frozen star would be overwhelming, enough to explain the intensely bright quasars that were being found by radio astronomers. At exactly the same time, an American astronomer at Cornell University, Edwin Salpeter, was coming to the same conclusion, that copious radio emissions could come from a massive object that weighed more than a million solar masses or, as he put it, "extremely massive objects of relatively small size."

Zel'dovich didn't stop there. With his young colleague Igor Novikov he applied his argument to binary systems such as a normal star circling a frozen star. They speculated that the immense gravitational pull of the frozen star would strip the outer layers of the normal star of all its gas and fuel. It is like, as Roger Penrose once put it, "having to drain . . . a bath the size of Loch Lomond through a normal size plughole." The forces that the gas would experience would be so tremendous that copious amounts of light at very high energy, known as x-rays, would be emitted. Look for the x-rays, Zel'dovich and his pupils told the world.

The name Schwarzschild was constantly popping up in scientific articles by astronomers and astrophysicists as the link between collapsed or frozen stars and quasars became more and more compelling. But, as Wheeler recalled years later, the name that he and his colleagues in the United States were using—"completely collapsed gravitational object"—was cumbersome, and "after you get around to saying that about ten times, you look desperately for something better." At a conference in Baltimore, in 1967, a member of the audience helped him out and proposed the term *black hole*. Wheeler adopted it, and the name stuck.

In 1969 one of Dennis Sciama's colleagues at Cambridge, Donald Lynden-Bell, stated in the introduction to one of his papers, "We would be wrong to conclude that such massive objects in spacetime should be unobservable, however. It is my thesis that we have been observing them indirectly for many years." He argued that massive black holes at the center of galaxies would suck in the surrounding material just as Penrose had described it, like water falling down a drain, gurgling around and around. The rotating gas around the hole would form a flat disk, just like Saturn's rings, and the whole system would be locked spinning on its axis. The nuclei of galaxies, fueled by these accretion disks, would then be veritable beacons of light, and Lynden-Bell could show how the energy was created and emitted. Martin Rees had also, with Dennis Sciama, set to work trying to build detailed models of quasars that could explain all the different strange properties — their size, their distances, how quickly they would flicker and pulsate, and what ranges of energy would be pumped out. Over the next few years, Rees, with Lynden-Bell and their students and postdocs at Cambridge, were able to come up with a beautiful, meticulous model of the fireworks surrounding quasars and radio sources. All the pieces were falling into place.

And then, finally, Zel'dovich and Novikov's x-rays started trickling in. Starting in the 1960s, a team led by the Italian physicist Riccardo Giaccone sent rockets up out of Earth's atmosphere where, for a few minutes, they would look for x-rays. They found them, bright spots of x-rays spread across the sky that outshone the planets in the solar system. In the early 1970s the *Uhuru* satellite was launched from a platform near Mombasa in Kenya with the sole goal of mapping out the x-ray sky. It was a resounding success, making exquisite measurements of over three hundred x-ray objects.

In the midst of the multiple sources that *Uhuru* measured was one object, Cygnus X-1, a particularly bright source lying in the constellation of Cygnus. It had first been seen in 1964 during one of the early rocket flights, but *Uhuru* found that its x-ray light flickered extraordinarily quickly, several times a second, a sure indication that it was an incredibly compact object. *Uhuru*'s measurements were rapidly followed by observations in radio frequency and optical frequencies that identified the smoking gun that Zel'dovich and Novikov had predicted — a star that is slowly being stripped of its envelope and gently wobbles as it is tugged about by an invisible, dense object with a mass of more than eight suns. There it was: the first evidence of a black hole; not certain, but highly probable. It was small, powerful, and invisible yet beaming out x-rays.

In the summer of 1972, Bryce and Cécile DeWitt organized a summer school at Les Houches in the French Alps. In attendance were the young relativists — trained by Sciama, Wheeler, and Zel'dovich — who had now become the world authorities: Brandon Carter and Stephen Hawking from Cambridge, Kip Thorne and his student James Bardeen along with Remo Ruffini from Caltech and Princeton, and Igor Novikov representing Moscow. They were the new prophets of black holes.

"The story of the phenomenal transformation of general relativity within little more than a decade, from a quiet backwater of research, harboring a handful of theorists, to a booming outpost attracting increasing numbers of highly talented young people . . . is by now familiar," the DeWitts wrote in the preface to the proceedings of the Les Houches meeting. "No single object or concept epitomizes more completely the present stage of evolution than the black holes." The meeting was the culmination of a decade of phenomenal discovery.

Einstein and Eddington had been profoundly mistaken. Even Wheeler had caved in and by 1967 had accepted that nature *didn't* abhor the singularities in general relativity. Schwarzschild's solution, discovered so long ago on the battlefields of the eastern front, and Kerr's solution, discovered in the heat of the Texas summer, were real and must exist in nature. They were the true endpoints of gravitational collapse. They were predicted by general relativity, inevitable and simple, and they could do wonderful things in nature: power quasars and strip stars of their envelopes. The radio sky, again and again, threw up tantalizing glimpses, and the x-ray mayhem that was being uncovered seemed to point to small dense objects. No measurement was yet definitive, but the real existence of black holes was becoming unavoidable. Bets were being made on which of the various strange beasts being observed in the sky could actually be black holes. They were almost a reality.

The group gathered at Les Houches had also, in the previous few years, realized that if black holes were to be found in nature, they *had to be* as mathematically simple as Schwarzschild's and Kerr's solutions. While Ezra ("Ted") Newman from Syracuse University had slightly extended Kerr's solution to include black holes that were electrically charged, the full black hole solution to Einstein's theory could be completely characterized in terms of just three numbers: its mass, how fast it was spinning, and how much charge it had. This was a startling result. Why couldn't a black hole have a bit more mass on one side, like a mountain on the surface of the Earth, compensated by less mass on the other side, like a valley? Or why couldn't it indent on one side while continuing to have the same mass? You could in fact imagine black holes with the same mass, spin, and charge all looking different, each having its own individual characteristics. But the math proved otherwise and showed resolutely that with general relativity such complications would quickly disappear. The hills would flatten out, the valleys would fill up, and the squashed areas would swell up. Black holes with the same mass, spin, and charge would all quickly settle down to look *exactly* the same as one another, completely indistinguishable. Wheeler described this uniform makeup by saying, "Black holes have no hair," and the proof became known as the "no-hair" theorem.

The Les Houches meeting showed what could happen when great minds tackle great problems. As Martin Rees recalls of that period, "There were three groups trying to understand black holes: Moscow, Cambridge, and Princeton. And I always felt there was a congenial atmosphere among them all." Indeed, during a time of tremendous isolation between the East and West, their collaborative meetings pushed the science forward. Kip Thorne and Stephen Hawking would visit Zel'dovich in Moscow and compare notes on accretion disks, gravitational collapse, and singularities. As important were the short and difficult trips to the West taken by the Soviet physicists. As Novikov recalls of his visit to one of the Texas Symposiums in 1967, this time in New York, "Despite our desperate efforts to gather maximum information and talk to as many colleagues as possible, we were physically unable to cover all that was of interest." Years later, at the Les Houches meeting in 1972, Novikov and Thorne would coauthor one of the papers on accretion disks.

In ten years, Einstein's theory of general relativity had been transformed. The Texas Symposium had become a regular gathering of many hundreds of astrophysicists, many of whom now considered themselves relativists. As Roger Penrose put it, "I saw black holes change from a piece of mathematics into something people actually believed in." The generation that came out of the Golden Age of General Relativity would be rewarded with prestigious positions at some of the top universities. In the United Kingdom, Martin Rees and Stephen Hawking would be given prestigious chairs at Cambridge, as would

Roger Penrose at Oxford. In the United States, Wheeler's students found themselves on the faculties of Caltech, Maryland, and a number of other top universities, as did Zel'dovich's offspring in the Soviet Union. All of this for their work on general relativity. It seemed that Einstein's theory had finally become part of mainstream physics in a truly spectacular way.

Chapter 9

Unification Woes

I N 1947, FRESH OUT OF graduate school, Bryce DeWitt met Wolfgang Pauli and told him he was working on quantizing the gravitational field. DeWitt couldn't understand why the two great theories of the twentieth century—quantum physics and general relativity—were kept at arm's length. "What is the gravitational field doing there, in such splendid isolation?" he wondered. "What if one simply dragged it forcibly into the mainstream of theoretical physics and quantized it?" Pauli hadn't been entirely supportive of DeWitt's plans. "That is a very important problem," he told him, "but it will take someone really smart." No one would deny DeWitt's considerable intelligence, but for more than half a century, general relativity would prove remarkably resistant to his efforts.

General relativity stood alone in its incompatibility with quantum physics. The ascent of the quantum after the Second World War led to a completely new and powerful theory that brought together all the forces with the fundamental constituents of matter as a simple, coherent whole—all the forces, that is, except gravity. Albert Einstein and Arthur Eddington had tried and failed for decades to come up with their own unified theories. Quantum theory was different. It was tested with staggering precision in gigantic collider experiments in Europe and the United States, a success story marrying beautiful mathematics and conceptual brilliance with real, down-to-earth measurements.

Despite its successes, there was one man who refused to cheer on the new postwar quantum physics. Paul Dirac thought the quantum theory of particles and forces was a sham and a piece of messy thinking. It performed a sleight of hand, sidestepping fundamental problems by making some infinite numbers magically disappear. Dirac was convinced this trickery was what prevented general relativity from joining in the full glory of the unification of *all* the forces.

There was something impenetrable about Paul Dirac, a tall, slim man who hardly said anything in polite company. When he did speak, his words were almost *too* precise and to the point. He would often come across as painfully shy and preferred to work on his own, obsessed with the mathematical beauty that he believed underpinned reality. His papers were mathematical gems with far-reaching real-world consequences. He originally trained as an engineer in Bristol but quickly established himself as one of the prophets of the new quantum when he came to Cambridge in his early twenties. He was rapidly made a fellow of St. John's College in Cambridge and soon afterward became the Lucasian Professor of Mathematics, a chair that had been filled by Isaac Newton in the seventeenth century. Cambridge gave him a sheltered existence where he could hide away yet also influence generations of physicists, among them some of the astrophysicists and relativists who came to reenergize general relativity in the 1960s. Both Fred Hoyle and Dennis Sciama had been his PhD students, and Roger Penrose had sat in on his lectures, marveling at their clarity and precision.

Ironically, it was Paul Dirac's own fundamental equation for the electron — the Dirac equation, as it became known — that took the first step on the path toward unification, bringing together Einstein's principle of special relativity and the foundations of quantum physics. The equations for quantum physics tell us how the quantum state of a system, such as an electron bound to a proton in a hydrogen atom, evolves with time. It makes a very clear distinction between space and time. Einstein's special relativity brings space and time together into one indivisible thing — spacetime. It also combines the laws of mechanics and the laws of light into a coherent framework. Paul Dirac was able to bring the laws of quantum physics into this same framework. With Dirac's equation, all of physics, including quantum physics, could obey the special principle of relativity.

Particles in the universe can be divided into two types — fermions and bosons. As a rule of thumb, the particles that make up stuff are mostly fermions, and the particles that carry the forces of nature are mostly bosons. Fermions include the building blocks of atoms, such as electrons, protons, and neutrons. As we have seen when looking at white dwarfs and neutrons stars, these particles have a bizarre quantum property that arises from the Pauli exclusion principle: No two fermions can occupy the same physical state. When squeezed into the same space, they push each other apart through quantum pressure. Fowler, Chandra, and Landau used this pressure to explain how white dwarfs and neutron stars sustained themselves below their critical mass. Unlike fermions, bosons do not satisfy the Pauli exclusion principle and can be compressed together at will. An example of a boson is the photon, the carrier of the electromagnetic force.

The equation that Dirac found describes the quantum physical behavior of an electron, a fermion, while also satisfying Einstein's special theory of relativity. It is an equation that describes the probabilities for finding an electron in any given position in space or with any given speed. Instead of singling out space, Dirac's equation is defined in all of spacetime in one coherent way, as special relativity demands. Dirac's equation contains a wealth of insights and information about the natural world and its fundamental particles. To his surprise, his equation also predicted the existence of antiparticles. An antiparticle has the same mass but the opposite charge of its corresponding particle. The antiparticle of an electron is called a positron. It looks exactly like an electron, but its charge is positive instead of negative. According to Dirac's equation, both electrons and positrons have to exist in nature. The equation also predicts that pairs of electrons and positrons can pop out of the vacuum, effectively created out of nothing. This was bizarre and difficult to understand, especially given that when Dirac first wrote down his equation no one had ever seen a positron. Dirac held back from claiming that positrons actually existed until, in 1932, they were detected in cosmic rays. Dirac won the Nobel Prize the following year.

When Dirac first proposed his equation, he started a revolution in the understanding of particles and forces in nature. If the quantum physics of the electron could be described in the same framework as the electromagnetic field — that is, obeying Einstein's special principle of relativity — why couldn't the electromagnetic field itself be quantized like the electron? Instead of just describing light waves, it should naturally describe photons as well, the quanta of light that Einstein had posited existed in 1905. A quantum theory of electrons *and* light, known as quantum electrodynamics, or QED for short, was the next step on the path to the unification of particles and forces. Developed by Richard Feynman, Julian Schwinger, and Sin-Itiro Tomonaga after the Second World War, it signaled a new way of studying quantum physics: quantizing particles (electrons) and forces (the electromagnetic field) in one coherent whole. QED was a phenomenal success and allowed its creators to predict the properties of electrons and electromagnetic fields with unprecedented precision, winning them the Nobel Prize as well.

While QED worked spectacularly well, Paul Dirac viewed it with tremendous disgust. For at the core of its success was a method of calculating that affronted his deep belief in mathematical simplicity and elegance. It went by the name of renormalization. To understand what renormalization means, we need to look at how physicists use QED to calculate the mass of an electron. The mass of an electron has been beautifully measured in laboratories and equals 9.1 tenths of a billionth of a billionth of a billionth of a gram — a very small number. However, applying the equations of QED gives you an infinite value for the electron's mass. This is because QED allows the creation and destruction of photons and short-lived pairs of electrons and positrons — the particles and antiparticles from Dirac's equation — effectively out of nothing. All these *virtual* particles popping out of the vacuum boost the self-energy and mass of the electron, ultimately making it infinite. And so QED, if applied injudiciously, leads to infinities all over the place and gives the wrong answer. But Feynman, Schwinger, and Tomonaga argued that since we *know* that the final mass of the electron from our observations is finite, we can simply take the calculated infinite result and "renormalize" it by replacing it with the known, measured value.

To the uncharitable observer it sounds like all renormalization does is throw away the infinities and arbitrarily replace them with finite values. Paul Dirac declared himself "very dissatisfied with the situation." As he argued, "This is just not sensible mathematics. Sensible mathematics involves neglecting a quantity when it is small — not neglecting it just because it is infinitely great and you do not want it!" It seemed like a messy piece of slightly magical thinking, but there was no denying that it worked spectacularly well.

QED was one step on the long path to unification, but from the 1930s to the 1960s it had become clear that there were two other forces, apart from the electromagnetic and gravitational forces, that also needed to be included in the ultimate framework. One was the weak force, proposed in the 1930s by the Italian physicist Enrico Fermi to explain a particular type of radioactivity known as beta decay. In beta decay a neutron transforms itself into a proton and spits out an electron in the process. Such a process is impossible to understand using electromagnetism, so Fermi conjured up a new force that would allow that transformation to happen. This new force acts only at very short distances, at internuclear separations, and is much weaker than electromagnetism; hence its name. The other force, the strong force, is what glues protons and neutrons together to form nuclei. It also binds the more fundamental particles, called quarks, that make up protons, neutrons, and a plethora of other particles. While it also has a very short range, it is much stronger than the weak force (hence the creative name). The challenge, just as James Clerk Maxwell had unified the electric and magnetic forces into a single electromagnetic force in the mid-nineteenth century, was to come up with a common way of dealing with all four fundamental forces: gravitational, electromagnetic, weak, and strong.

Throughout the 1950s and 1960s both the strong and weak forces were systematically unpeeled and studied in detail. As they became better understood, a mathematical similarity began to emerge between them and the electromagnetic force, suggesting there might be *one* unified force that manifests as one of the three different forces depending on the situation. By the late 1960s, Steven Weinberg of MIT, Sheldon Glashow of Harvard, and Abdus Salam of Imperial College in London had proposed a new way of packing at least two of the forces, the electromagnetic and weak forces, together into one electroweak force. The strong force couldn't yet be brought into the mix but looked so similar to the other forces that there was a belief that it should be possible to come up with a "grand unified theory" of the electromagnetic, weak, and strong forces. In the 1970s, the electroweak theory and the theory of the strong force were shown to be renormalizable, just like QED. All the pesky infinities that arose in their calculations could be replaced by known values, making the theories eminently predictable. The combination of the electroweak and strong theories became known as the standard model and made accurate predictions that were confirmed in laboratories like the gigantic particle accelerator at CERN in Geneva, Switzerland. This almost completely unified, yet powerful and predictive *quantum* theory of the three forces — electromagnetic, weak, and strong — was universally accepted.

By all, that is, except Paul Dirac. Although he was impressed with the younger generation that had put together the standard model and marveled at some of the mathematics that had been used, he repeatedly railed against the infinities and what he considered to be the nefarious trick of renormalization. In the few public lectures he gave in which he deigned to mention the standard model, he chided his colleagues for not trying harder to find a better theory with no infinities. Toward the end of his career at Cambridge, Dirac became more and more isolated. He stubbornly rejected the developments in quantum physics. Despite his craving for privacy, he felt ignored by the rest of the physics world, which had embraced QED and saw him as a figure of the past. So he withdrew, keeping to his study at St. John's College and avoiding the department where he held his professorship, paying no attention to the great discoveries in general relativity that were coming from Dennis Sciama, Stephen Hawking, Martin Rees, and their collaborators. As one of their contemporaries at Cambridge recalls, "Dirac was this ghost we rarely saw and never spoke to." He retired from his position as the Lucasian Professor in 1969 and moved to

Florida to take up a professorship there. In his final years he wouldn't have been surprised to see general relativity refuse to bow to the techniques of renormalization.

Bryce DeWitt had no idea what a struggle his pursuit of a quantum theory of gravity would be. While working with Julian Schwinger at Harvard, he had witnessed the birth of QED firsthand. When he decided to tackle gravity, DeWitt chose to treat it just like electromagnetism and tried to reproduce the successes of QED. There were similarities between electromagnetism and gravity: both were long-range forces that could extend over large distances. In QED, the transmission of electromagnetic force could be described as being carried by a massless particle, the photon. You can view electromagnetism as a sea of photons zipping back and forth between charged particles, like electrons and protons, pushing them apart or pulling them together, depending on their relative charges. DeWitt approached a quantum theory of gravity in an analogous way, replacing the photon with another massless particle, the graviton. These gravitons would bounce back and forth between massive particles, pulling them together to create what we call the gravitational force. This approach abandoned all the beautiful ideas of geometry. While gravity was still described in terms of Einstein's equations, DeWitt chose to think of it as just another force, bringing to bear all the techniques of QED.

For the next twenty years, DeWitt tried to figure out how to quantize the graviton, but he found it a gargantuan challenge. Once again Einstein's field equations were simply too unwieldy and entangled to be dealt with easily. He watched as the theory of the other forces developed and saw the similarities in the difficulties. But while the problems with unifying the strong, weak, and electromagnetic forces seemed to fall away, general relativity was obstinate, unwilling to be shoehorned into the same set of quantum rules that seemed to apply to the other three forces. Through his battle, DeWitt was not alone: Matvei Bronstein, Paul Dirac, Richard Feynman, Wolfgang Pauli, and Werner Heisenberg had all had a go at quantizing the graviton at some point before him. Steven Weinberg and Abdus Salam, the architects of the successful model of the electroweak force, attempted to apply the techniques that they had developed for the standard model, but they too found that gravity was too difficult.

As DeWitt labored on, grappling with the graviton and trying to quantize it, isolated pockets of interest in his work developed. John Wheeler cheered him on and set his students working on it, as did the Pakistani physicist Abdus Salam, Dennis Sciama in Oxford, and Stanley Deser, based in Boston. But in general, reactions to work on quantum gravity were mixed and often cool. Michael Duff, a former student of Salam, recalls presenting his results on quantum gravity at a conference in Cargèse, Corsica, and being "greeted with hoots of derision." A student of Dennis Sciama named Philip Candelas, who was working on quantum properties of fields living on spacetimes with different geometries, heard that members of the faculty of physics at Oxford were muttering that he "wasn't doing physics." Quantum gravity was still too unformed compared to the work on quantizing the other forces. To many, it was perceived as a waste of time.

In February 1974, the United Kingdom was at a standstill. The price of oil had shot up, a succession of ineffectual governments had been trying to stem the rise of inflation, and the country was hamstrung by industrial strife. Every now and then the working week was shortened to three days to save energy, and rolling power cuts meant that evening meals were often eaten by candlelight. It was during these dark days that a meeting was convened to take stock of the progress in quantizing gravity, almost twenty-five years after DeWitt first set to work. Despite the somber economic climate, euphoria reigned at the start of the Oxford Symposium on Quantum Gravity. The predictions of the standard model of particle physics developed by Glashow, Weinberg, and Salam were being spectacularly confirmed at the massive particle accelerator at CERN. Surely quantum gravity would have to follow close behind.

Yet, as the speakers stood up and presented hints of solutions and ideas, again and again, the same problem seemed to scupper the most promising and popular route for quantizing gravity. DeWitt's approach of forgetting about geometry and thinking of gravity simply as a force was not working. The organizers, paraphrasing Wolfgang Pauli, fretted, "What God hath torn asunder, let no man join." The problem was that general relativity was not like QED and the standard model. With QED and the standard model it was always possible to renormalize all the masses and charges of the fundamental particles and get rid of the infinities that cropped up to get sensible results. But if the same tricks and techniques were applied to general relativity, the whole thing fell apart. Infinities kept on cropping up that refused to be renormalized. Tuck them away in one part of the theory and they would stick out in another part, and renormalizing the whole theory in one fell swoop proved impossible. Gravity, as described by general relativity, seemed far too entangled and different to be repackaged and fixed like the other forces. At the symposium, Mike Duff said ominously in the conclusion to his talk, "It appears that the odds are stacked against us, and only a miracle could save us from non-renormalizability."

Quantum gravity had hit a dead end, and general relativity refused to join the other forces in one, unified picture. As a *Nature* article on the symposium glumly noted, "The presentation of technical results by M. Duff only served to confirm the extraordinary lengths which are necessary to make even minor progress." This failure was all the more galling given that there had been such tremendous progress in relativistic astrophysics, black holes, and cosmology in the previous years, not to mention the spectacular success in the standard model of particle physics.

The Oxford symposium seemed like an admission of defeat, except for one surprising talk by the Cambridge physicist Stephen Hawking on black holes and quantum physics. In his talk, Hawking showed that there was a sweet spot where quantum physics and general relativity could be brought together. Furthermore, he claimed he could prove that black holes weren't in fact black but shone with an incredibly dim light. It was an outlandish claim that would transform quantum gravity for the next four decades.

By the early 1970s, Stephen Hawking was already a fixture on the Cambridge scene, working at the Department of Applied Mathematics and Theoretical Physics, or DAMTP for short. At only thirty, he had already made a name for himself in general relativity. Coming out of Dennis Sciama's stable of students, Hawking had worked with Roger

Penrose to show that singularities had to exist in the very beginning of time. In the early 1970s he had turned his attention from cosmology to black holes and, with Brandon Carter and Werner Israel, had proved definitively that black holes have no hair: they lose any memory of how they were formed, and black holes with the same mass, spin, and charge all look exactly alike. He had also obtained an intriguing result about the sizes of black holes. If you took two black holes and merged them together, he found, the area of the Schwarzschild surface, or event horizon, of the final black hole had to be greater than or equal to the sum of the area of the original black holes. In practice, this meant that if you summed up the total area of black holes before and after *any* physical event, it *always* increased.

Hawking did all this work as Lou Gehrig's disease claimed his body. Throughout the late sixties, he walked through the corridors at DAMTP with a cane, leaning against the wall for support, but he slowly and steadily became unable to move unaided. As his ability to write and draw, essential tools in the arsenal of a theoretical physicist, dwindled away, he developed a formidable capacity to think things through at length, allowing him to tackle deep issues in general relativity and quantum theory.

One might say Hawking's great discovery was driven by his annoyance at a result put forward by a young Israeli PhD student of John Wheeler named Jacob Bekenstein. Bekenstein wanted to reconcile black holes with the second law of thermodynamics. To do so, he used one of Hawking's results to come up with a completely ludicrous claim about black holes. To Hawking, the claim was entirely too speculative and simply wrong.

To understand Bekenstein's claim, we need to take a quick detour into thermodynamics, the branch of physics that studies heat, work, and energy. The second law of thermodynamics (there are four in total) states that the entropy, or level of disorder, of a system always increases. Consider the classic example of a simple thermodynamic system: a box containing gas molecules. If the molecules are all at rest, neatly packed away in one corner, the system has low entropy — there is very little disorder. There is also no way the stationary particles will collide with the sides of the box and heat it up, so the system has a low temperature. Now imagine that the molecules begin to move. They roam freely throughout the box and spread out randomly, shifting the system to a high-entropy state. That is, the distribution of molecules inside the box becomes more disordered. As they move around, they collide with the walls of the box and transfer some of their energy to it, heating it up and increasing its temperature. The faster the molecules move, the quicker they randomize, and the quicker the entropy goes up until it reaches its maximum. Indeed, the quicker the molecules move around, the less likely it is that they will all coalesce into a peaceful, ordered state of low entropy. But not only that, faster molecules also transfer more heat to the walls of the box, increasing the temperature of the system even more. This shows us two things: the box tends toward a high-entropy state, as the second law of thermodynamics states, and with entropy comes temperature.

Bekenstein wanted to address the paradox of what would happen if you threw a box of stuff into a black hole. The box could contain anything: encyclopedias, hydrogen gas, a lump of iron. To keep it simple, let's consider our box of gas. The box will disappear into the hole and very rapidly the no-hair theorem will kick in. After the event, there will be no way of knowing what originally fell in. All information about the box will be lost. But if this is so, all the disorder of the gas in the box — all that entropy — has also disappeared, and the total entropy of the universe has gone down. Black holes appear to violate the second law of thermodynamics.

The way that Bekenstein found to salvage the second law of thermodynamics was to use Hawking's result. If you throw stuff into a black hole, the area of the event horizon never decreases — it either stays the same or increases. And so Bekenstein concluded that if the second law of thermodynamics is to be satisfied in the universe, black holes *must* have entropy, directly related to the surface area of the event horizon. The increase in the area of the black hole would more than compensate for the loss of disorder, sucked in behind the event horizon, and the entropy of the universe could never decrease. Yet, if Bekenstein pushed his solution of the paradox to its ultimate consequences, he came up with a bizarre result. If a black hole has entropy, then, just like the box of gas molecules, it should also have a temperature. Even Bekenstein felt he was going too far and wrote in his paper, "We emphasize that one should not regard T as the temperature of the black hole; such an identification can easily lead to all sort of paradoxes, and is thus not useful."

Despite Bekenstein's reservations, Hawking found his claim galling. According to the laws of thermodynamics, there is no way to increase the entropy of a black hole without causing it to radiate heat in some way. For Hawking, this was going too far. To him, it was obvious that black holes were black: things could fall into black holes, but they definitely couldn't come out. The fact that the overall area of black holes couldn't decrease, as he himself had shown, might look like entropy, but it wasn't *really* entropy — entropy was just a useful analogy for explaining the behavior.

But there were hints that Bekenstein might be right and Hawking wrong. For a start, in 1969 Roger Penrose found that a spinning black hole, described by Kerr's solution, could emit energy. Imagine a fast-moving particle traveling at close to the speed of light as it falls into the orbit of a Kerr black hole. If it decays into two particles, one of which is sucked into the event horizon, the remaining particle can be sped up and thrown out with more energy than went in, conserving the total energy of the system, and the universe. With this odd process, known as Penrose superradiance, black holes are effectively emitting energy as if they are shining in some bizarre way. But there were more ideas floating around. In 1973 Stephen Hawking visited Yakov Zel'dovich and his young colleague Alexei Starobinsky and learned that they had worked out what would happen to a Kerr black hole: it would strip away the quantum vacuum that surrounded it, using its energy to emit energy and indeed radiate.

Hawking decided to use quantum physics to think about particles close to the event horizon of a black hole, where strange things could happen. What he found there was strange indeed. Quantum physics allows pairs of particles and antiparticles to form out of the vacuum. In ordinary circumstances these particles are created and then, just as quickly, collide with each other and are annihilated, completely disappearing. But, as Hawking found, the situation is very different near the event horizon: some of the antiparticles will be sucked into the black hole while the particles remain. This process will happen again and again, and as the antiparticles are sucked in, the black hole will, slowly and surely, emit a stream of energetic particles. Hawking worked out the details of what would happen if the particles were massless, like photons. And he found that, observed from afar, the black hole would shine with an incredibly low brightness, very similar to a dim star. And just like a star, our sun, for example, it would be possible to assign it a temperature. By looking at the light our sun emits, we can measure its surface temperature to be about 6,000 degrees Kelvin. In other words, *because* of quantum physics, Hawking had found that the black holes predicted by general relativity emitted light and had a temperature.

It was a remarkably clean and unambiguous mathematical result with far-reaching consequences. Hawking's calculation was able to show that the temperature with which a black hole shines is inversely proportional to its mass. So, for example, a black hole with the mass of the sun would have a temperature of a billionth of a Kelvin, and a black hole with the mass of the moon would have a temperature of about 6 degrees Kelvin. As the black hole shines, it sheds some of its mass. This process happens incredibly slowly. A black hole with the mass of the sun would take an inordinately long time to shed all its mass, or "evaporate," as Hawking described it. But much smaller black holes could evaporate much more quickly. So, for example, a black hole with a mass of about a trillion kilograms (a small black hole from an astrophysical point of view) would evaporate within the lifetime of the universe, releasing a burst of energy in the last tenth of a second. As Hawking described it, it would be "a fairly small explosion by astronomical standards but it is equivalent to about 1 million 1-Mton hydrogen bombs." Hawking called his paper, which he would end up publishing in *Nature,* "Black Hole Explosions?"

When Stephen Hawking presented his talk at the Oxford symposium, he sat awkwardly in a wheelchair at the front of the auditorium. He had something groundbreaking to say, and he spoke clearly and purposefully, explaining his calculations to the gathered audience. When he finished, he was met by near silence. As Philip Candelas, a student of Dennis Sciama at the time, recalls, "People treated Hawking with great respect but no one really understood what he was saying." As Hawking himself later recalled, "I was greeted with general incredulity. . . . The chairman of the session . . . claimed it was all nonsense." In the review of the Oxford symposium in *Nature,* it was acknowledged that "the main attraction of the conference was a presentation by the indefatigable S. Hawking," yet the author of the review was skeptical about his prediction of exploding black holes, writing, "Exciting though this prospect may be, no plausible physical mechanism could be discerned which might lead to such a dramatic effect."

It would take some time for Hawking's discovery to sink in, but a few people immediately realized the significance of what he had done. Dennis Sciama referred to Hawking's paper as "one of the most beautiful in the history of physics" and immediately set some of his students to work on pushing it further. John Wheeler described Hawking's result as "like candy rolling on the tongue." Bryce DeWitt set about rederiving Hawking's result his own way and wrote a review of black hole radiation that would convince a whole new group of people.

Hawking's calculation of black hole radiation wasn't quantum gravity. It didn't involve quantizing the gravitational field by working out the rules and processes that gravitons would be subjected to, as DeWitt and so many had been trying and failing to do. But it did successfully mix the quantum and general relativity to give an interesting hard result, something that quantum gravity, if it ever came to fruition, might refer to and explain in more detail. And so, over the next few years, black hole radiation brought new hope to the impossible challenge of quantizing gravity. Hawking firmly trained his sights on quantizing not only objects within spacetime but spacetime itself. Training a new set of students to work on his program, Hawking would remain intensely focused on quantum gravity for the next forty years. It was fitting, then, that ten years after Paul Dirac retired from the Lucasian Professorship at DAMTP, Stephen Hawking was appointed to it, a position he ended up holding for over twenty-five years.

When John Wheeler was asked by a young student how he could best be prepared for working on quantum gravity — would it be better to be an expert in general relativity or in quantum physics? — he replied that it would probably be better if the student worked on something else altogether. It was wise advice. Stubborn infinities continued to thwart every attempt at quantizing general relativity, and it seemed that any endeavor in the quest for quantum gravity was destined for failure.

Yet it was also true, as Hawking had shown with his spectacular result, that when general relativity and quantum physics did meet, unexpected things happened. Black holes had entropy and emitted heat, which went against the idea relativists had of black holes being, well, black. But Bekenstein's and Hawking's calculations also seemed to shed new light on the quantum, to which general relativity seemed to do odd things. In a usual, run-of-the-mill physical system, like a box of gas, the entropy is related to volume. The more volume there is, the more possible ways there are to randomize things and create disorder, the hallmark of entropy. All that randomness, that disorder, is stored away *in* the box. The direct relation between entropy and volume is part and parcel of textbook thermodynamics. But what Bekenstein and Hawking found, as we saw, is that the entropy of the black hole is related to the *area* of its surface and not to the volume it takes up in space. That's like our box full of gas particles somehow storing its entropy in the walls of the box instead of in the random movements of the gas particles within. How do we store entropy on a black hole's surface, which, as we know, should be simple and hairless, just uniformly emitting light through Hawking radiation?

Intractable and inscrutable, with all of the new mind-boggling results in black holes, quantum gravity had become the ultimate challenge for clever young physicists. Yet, while quantum gravity became a veritable battleground of ideas that would play out over the following decades, another battle was taking place in general relativity. Instead of thought experiments and clever mathematics, it involved instruments and detectors trying to measure elusive waves in the fabric of spacetime emanating from colliding black holes.

Chapter 10

Seeing Gravity

JOSEPH WEBER WAS once heralded as the first observer of gravitational waves. He created the field of gravitational wave experiments almost single-handedly. In the late 1960s and early 1970s, Weber's results were celebrated as major accomplishments for relativity. But by 1991, he had been brought low. As he told his local newspaper that year, "We're number one in the field, but I haven't gotten any funding since 1987."

On the face of it, Joe Weber's situation seemed bizarrely unfair. At the height of his career, his results were discussed at all the major conferences of general relativity alongside neutron stars, quasars, the hot Big Bang, and radiating black holes. They were the subject of countless papers trying to explain them. Weber was a shoo-in for a Nobel Prize. And then, just as quickly as he had risen to prominence, Weber was cast out into the hinterland of academia. Shunned by his colleagues, rejected by the funding agencies, unable to publish in any of the mainstream journals, Weber was condemned to a long and lonely scientific death, an odd and uncomfortable footnote in the history of general relativity. Some would even say that it was only after Weber's fall that the real quest for gravitational waves began.

Gravitational waves are to gravity what electromagnetic waves are to electricity and magnetism. When James Clerk Maxwell showed that electricity and magnetism could be unified into one overarching theory, electromagnetism, he set the foundations for Heinrich Hertz to show that there would be electromagnetic waves that would oscillate at a range of frequencies. At visible frequencies, these waves would be the light that our eyes are so attuned to picking up and interpreting. At longer frequencies, these would be the radio waves that bombard our radio receivers, transmit the wireless information to and from our laptops, and allow us to see the immensely energetic quasars out in the far recesses of the universe.

Within months of coming up with general relativity, Albert Einstein had shown that, just like electromagnetism, in his new theory, spacetime should contain waves. The waves would be ripples in space and time themselves. Spacetime acts sort of like a pond; when you throw in a pebble, it sends out ripples that propagate from one end to the other. Just like electromagnetic waves and the ripples of water in a pond, gravitational waves can carry energy from one place to another.

Unlike electromagnetic waves, gravitational waves have proved incredibly difficult to find. They are very inefficient at carrying energy out of a gravitating system. As the Earth orbits around the sun at a distance of 150 million kilometers, it slowly loses energy through gravitational waves and drifts closer to the sun, but the distance between the Earth and the sun shrinks at a minuscule rate, about the width of a proton per day. This means that during its whole lifetime, the Earth will drift closer to the sun by a mere *millimeter*. Even if something is large enough to generate a copious amount of gravitational waves, those waves become the faintest whispers when they travel through spacetime. Spacetime is actually less like a pond of water and more like an incredibly dense sheet of steel that barely trembles at the hardest of kicks.

Gravitational waves were hard for other physicists to stomach. For almost half a century after Einstein argued that they existed, many refused to believe they were real. They were seen as yet another mathematical oddity that could be explained away with a deeper understanding of Einstein's general theory of relativity. Arthur Eddington, for one, staunchly rejected the existence of gravitational waves. Having repeated Einstein's calculation in which he worked out how gravitational waves would appear in general relativity, he went on to argue that they were an artifact of how you chose to describe space and time. They arose because of a mistake, an ambiguity in labeling positions in space and time, and could be done away with completely. These waves weren't real waves, and unlike electromagnetic waves that traveled at the speed of light, Eddington dismissed gravitational waves for traveling at the "speed of thought." In a surprising turn of events, Einstein himself decided that he had been mistaken in his original calculation, and in 1936 he submitted a paper along with one of his young assistants, Nathan Rosen, to the *Physical Review* in which they argued that gravitational waves simply couldn't exist.

Hermann Bondi made the most compelling case for gravitational waves at the Chapel Hill meeting in 1957. Bondi, then leading a relativity group at King's College London, presented a simple thought experiment: Take a rod and thread it through two rings a small distance apart from each other. Tighten the rings ever so slightly so that they can still move but rub against the rod. If a gravitational wave passes through, it will barely affect the rod itself. The rod will be too stiff to sense anything much. But the rings will be dragged up and down on the rod, like buoys in the sea being tossed about by the waves. They will move back and forth, coming close and moving apart as the wave flies through, and in doing so they will rub against the rod and heat it up, giving it energy. Given that the only place the energy could come from is the gravitational waves, the waves must carry energy. Bondi's argument was simple and effective. Richard Feynman, who was also attending the meeting, presented a similar line of reasoning, and the majority of the participants were convinced. Gravitational waves were out there, ready to be discovered. Joe Weber had been at Chapel Hill, mesmerized by the discussions. Bondi, Feynman, and all the other participants could sit around discussing the reality of gravitational waves, but he would actually go out and look for them.

Weber was just the sort of person who would attempt the impossible. An obsessive tinkerer, he had learned to fix radios to make money as teenager. An artistic visionary, constantly pushing technology beyond what was thought feasible, he would design and build experiments with the barest resources and then use them to probe the outer edges of the physical world. His drive infected all aspects of his life; he ran three miles every morning and worked a full day until he was in his late seventies.

Weber had trained at the United States Naval Academy as an electrical engineer and commanded a ship during the Second World War. Because of his expertise in electronics and radio he was asked to lead the navy's electronic countermeasures program. When he came out of the war, he became a professor of electrical engineering at the University of Maryland, where he decided to switch fields, studying for a PhD in physics.

In the mid-1950s, Weber became interested in gravity. John Wheeler had stepped in and encouraged Weber to take the plunge, bringing him to Europe for a year to think about the new frontier of general relativity. When Weber returned, he was ready to start designing and building an instrument. As he gradually immersed himself in the task of recording gravitational waves, he sketched out various possibilities, filling up notebooks with designs for contraptions. One method particularly took his fancy. The idea was simple: Build big, heavy cylinders of aluminum and suspend them from the ceiling. Strapped around the belly of each cylinder would be a set of incredibly sensitive detectors that would send an electrical pulse to a recorder if the cylinder vibrated. Anything could set it off—a phone ringing, a car trundling by, a slamming door. So Weber had to isolate the cylinders as much as possible, eliminating all possible sources of tremors and jerks.

When Weber finally turned on his cylinders, or Weber bars as they became known, he immediately began to pick up tremors. The bars vibrated, and once all the known disturbances had been eliminated, a few were left over: little blips of what just might be gravitational radiation. There was something odd about the blips, though. If they were truly gravitational radiation, they must have come from such an explosive event that it would have been observed through telescopes.

The signal was too strong to be gravitational radiation. Weber had to improve his kit.

To be absolutely sure that any tremor in the cylinders came from a gravitational wave passing through, Weber placed one of his four bars at the Argonne National Laboratory, almost a thousand kilometers away from his laboratory at the University of Maryland. If cylinders at *both* places trembled at the same time, it would be a strong sign that they were being sprayed by gravitational waves coming from outer space. Weber would compare the readings of the detectors on each one of his bars. If a reading shot up on more than one bar *at the same time,* it would be more likely that the source of the disturbance was the same external thing — a gravitational wave — that had shaken both of the bars, and not just some randomly coordinated jiggle in each of the bars themselves. He would look for these "coincidences," as he called them. Once again, Weber turned on his machine and waited.

By 1969, after working on his experiment for over a decade, Weber had something to show the world: a handful of coincident tremors not only between the ANL and the University of Maryland cylinders but between all *four* of his cylinders. It was too much of a coincidence to be random. They must have been sensing something in unison. There were no earthquakes, nor was there any strange electromagnetic storm to which he could attribute the phenomenon. Weber appeared to have discovered gravitational waves.

Over the next few years, Joseph Weber perfected his experiment, making sure that he was not simply finding what he wanted to find. The tremors in the bars were few and far between and were buried in the noise of the experiment. The bars would jiggle simply because of their own heat, as the atoms and molecules within them vibrated back and forth, and if you weren't careful, your eyes would pick up patterns where there were none. To get around this, Weber developed a computer program that would pick out the tremors and identify the coincidences automatically. He also decided to introduce a slight delay in recording the signal of one of the cylinders and then compare it with the other cylinders. If the coincidence were indeed true, the signal from one cylinder would arrive at the other time-delayed cylinder *after* the coincidence had actually happened — the number of coincidences would have to go down when comparing the records of the two cylinders. And, indeed, the number of coincidences fell.

By 1970, Weber had been running his experiment long enough that he was able to pinpoint the direction of the gravitational radiation that his instrument was picking up. It seemed to be emanating from the center of the galaxy, which he saw as a good thing. As he wrote in his paper, "A good feature is the fact that [10 billion] solar masses are there and it is reasonable to find the source to be the region of the sky containing most of the mass of the galaxy."

As Weber became more convinced that he was actually detecting gravitational waves with his experiment, the rest of the world began paying attention. His discovery had caught everyone by surprise. Such a straightforward detection of gravitational waves was unexpected, yet there was no reason, a priori, to doubt his findings. Weber's results were being brought up repeatedly by the relativists as they tried to figure out what they meant. Roger Penrose calculated what would happen if two gravitational waves collided with each other — could the final result be so explosive that it would trigger Weber's machine? Stephen Hawking worked out his own thought experiment of throwing black holes at each other, hoping that they would send out a burst of gravitational radiation that could explain Weber's detection. And throughout those early years, Weber's fame continued to spread. He was interviewed for *Time* magazine, and his work was featured in the *New York Times* and countless other newspapers in the United States and Europe. The results kept on pouring in.

Weber's results were amazing, and they seemed almost too good to be true. Weber appeared to have found an unbelievable source of gravitational radiation, far bigger than anyone had ever thought possible. For however sophisticated Weber's bars were and however refined the detectors he had glued to them, they weren't *that* sensitive. To actually get a detectable tremble, Weber's bars would have to be shaken by incredibly powerful gravitational waves, real behemoths traveling toward the Earth.

That was a problem, for even though the presumed gravitational waves came from the center of the galaxy, where there was a lot of stuff ready to implode, collide, and stir up spacetime, that was over twenty thousand light-years away from Earth. If indeed there was a beacon of gravitational waves lurking at the heart of the Milky Way, the waves it emitted would have been diluted in the intervening space into almost nothing by the time they reached the Earth. In fact, as Weber pointed out, the amount of energy in the gravitational waves he was detecting was equivalent to a thousand stars the size of the sun being destroyed at the center of the galaxy each year, a truly colossal amount.

Martin Rees at Cambridge was skeptical of Weber's results from the beginning. With his former PhD adviser, Dennis Sciama, and George Field from Harvard University, Rees worked out how much energy could be flooding out of the center of the galaxy in the form of gravitational waves. Rees and his collaborators found that, at most, two hundred stars the size of the sun could be destroyed each year to give rise to the gravitational waves. Any more than that, and the galaxy would have to be inflating, which they could verify was not the case by looking at the motion of nearby stars. Their calculation was approximate, so they were careful about their conclusions. In their paper they claimed, "Since the high rate of mass loss indicated by Weber's experiments is not ruled out by direct astronomical considerations discussed here, it would be clearly desirable for these experiments to be repeated by other workers." Weber was undaunted, for it was a *theoretical* argument that Rees, Field, and Sciama were putting forward. Maybe the theory was wrong, but his experiments were definitely right.

Following Weber's lead, in Moscow, Glasgow, Munich, Bell Labs, Stanford, and Tokyo new sets of experiments were being built. Some were exact copies of Weber's, and all of them were in one way or another inspired by Weber's original raft of designs. As they were gradually switched on, results started to trickle in, and a common pattern began to emerge; apart from a few events in the detector at Munich, none of them seemed to find the copious amounts of coincidences that Weber was finding with his apparatus. They simply weren't there. Weber was unfazed. He had a ten-year head start thinking about these experiments, and it was clear to him that all the other experiments were much less sensitive than his, so there was no surprise that there was no signal. If they wanted to criticize his results, they should build a detector *exactly* like the one he had built, a "carbon copy." Then they could talk. Several of the experimenters, including those in Glasgow and at Bell Labs in Holmdel, rebutted that the experiments they had built *were* carbon copies, and they still weren't seeing anything like what Weber was finding. Again Weber had an excuse: their copies simply weren't good enough.

But there was something troubling about Weber's own experiment. For a start, his bars weren't necessarily more sensitive than all the others. In such a nascent field, it wasn't yet clear how to determine the sensitivity of the experiments. But more worrying was the fact that Weber was prone to making mistakes but *still* found coincidences. For a start, he had claimed that the gravitational waves he was measuring came from the center of the galaxy. He concluded this by realizing that the tremors were mostly happening in clusters of events every twenty-four hours, when the bars were oriented toward the center of the galaxy. But Weber had missed an important point: gravitational waves would have no problem passing *through* Earth. So if the bars were aligned with the center of the galaxy but on the opposite side of the Earth, he should expect to find the same amount of coincidences. The clusters should happen every twelve hours, not every twenty-four hours as Weber had found. When Weber realized he had made a mistake, he went away and reanalyzed the data to find that, indeed, there was a twelve-hour cycle of coincidence that he hadn't picked up in his initial analysis. He seemed to find what he wanted to, once he knew what he was looking for. Bernard Schutz, a young relativist at the time, recalls that "people were very suspicious. He wasn't releasing his data so we could all have a look at it, yet he seemed to find whatever he wanted."

An even more glaring problem came up when Weber joined forces with another experimental team at the University of Rochester. As with his own cylinders, when Weber compared the tremors from the Maryland cylinders with the Rochester ones, he found a bundle of coincidences, vibrations that seemed to happen at exactly the same time in both places, a sure sign of gravitational waves. It turns out that Weber had misunderstood the way that the Rochester team had logged the time of each event, and the coincidences that Weber found in fact occurred with a four-hour time difference. Once the time delay was corrected, Weber analyzed the data again, and, again, he found coincidences.

Weber's discovery seemed to be impervious to mistakes and miscalculations. He could find coincidences anywhere. And coincidences meant gravitational waves. Weber's unwavering ability to bypass errors had a devastating impact on his reputation. He wasn't helped by the fact that no one else could reproduce his results. One respected experimentalist, Richard Garwin, wrote an article in *Physics Today* with the title "Detection of Gravity Waves Challenged" that systematically tore apart Weber's own data analysis and experiment, stating categorically that Weber's coincidences "*did not* result from gravity waves and furthermore *could not* have resulted from gravity waves." The community of relativists, en masse, turned their backs on Weber. Though he had once produced a stream of high-profile papers, Weber's publication rate plummeted. His funding dried up as more and more of his colleagues refused to support his prolific experiments. By the late 1970s, Weber had been cast out from the physics establishment.

Weber's experiments may have been discredited, but his results had set something much, much greater into motion. A new field was born out of the turmoil. Astronomers had realized that instead of capturing electromagnetic waves, such as light waves, radio waves, or x-rays, they could use gravitational waves as a new way of looking at the universe. Better, they could see *with* gravitational waves and look at things out in the farther recesses of spacetime that they couldn't see when they used conventional telescopes. Optical, radio, and x-ray astronomy would be joined by gravitational wave astronomy.

In 1974, two American astrophysicists, Joe Taylor and Russell Hulse, discovered not one but two neutron stars orbiting each other in a very tight orbit. One of the neutron stars was a pulsar, emitting bursts of light every few thousandths of a second, and could be easily followed as it orbited its silent companion. As these neutron stars orbited each other, Taylor and Hulse could measure their positions incredibly accurately. They had discovered a new, perfect laboratory for general relativity. Einstein had claimed that two objects orbiting each other would lose energy to the surrounding spacetime and their orbits would shrink until, ultimately, they would fall into each other. Although he abandoned this claim in later life, the calculation was there, ready to be tested. And the Hulse and Taylor millisecond pulsar could be used for exactly that.

In 1978, at the ninth Texas Symposium, held in Munich, Joe Taylor announced a new result. Having followed the millisecond pulsar for four years, he could safely say that the orbit was shrinking and that it was doing so as Einstein had predicted. As the two neutron stars orbited each other, they were losing energy through gravitational radiation. The evidence for gravitational radiation was indirect, but it was definitely there. It agreed beautifully with theory, and the measurements were clean and unambiguous. Gravitational waves were real.

Out of the debris of Weber's detection, a new field of experimental science was emerging. Different groups throughout the world were building their own detectors. Some were tweaking Weber's original design, cooling the cylinders dramatically so that they wouldn't vibrate at room temperature. Others changed the shape of the receptors, building spheres that would be sensitive to waves coming from all directions. But the signals they were looking for were so minute and so elusive that a bigger and better receptor was needed, one that would have the obscene sensitivity required to pick up ripples in spacetime. There was one approach that stood out from the others, hugely more powerful but also hugely more expensive: laser interferometry.

A laser interferometer makes use of the best tools of modern physics. For a start, it uses a laser beam, an incredibly focused ray of light that has been amplified and focused onto a very tiny target. Properly done, you can shine a laser for miles and it will hit its target, lighting up a pencil tip. Joe Weber had actually been one of the first people to come up with the concept of the laser, in his life before gravitational waves. He did so at the same time as Charles Townes at Columbia University but was never fully credited for his contribution and wasn't one of the awardees honored with the Nobel Prize in 1964 for its discovery.

Laser interferometry also makes use of another property of light, the fact that it behaves like a wave. Imagine waves in the ocean. When two waves with the exact same wavelength meet, they interfere. This means that if the waves meet when they are both at a crest, they add up constructively, and the resulting wave will have a much higher crest (and much deeper trough). But if they meet and one of them is at a crest and the other one at a trough, they will cancel each other out and interfere destructively. There is, of course, a whole range of behaviors between these two extremes.

These two properties of laser light can be used to detect minute motions of objects that have been affected by gravitational waves. The instruction manual is as follows: Suspend two masses at a distance from each other and shine a laser beam onto each of them. Each of the beams will reflect off the masses and interfere with the other; the resulting interference pattern will depend on the wavelength and the exact distance traveled. If one of the masses shifts ever so slightly, the interference pattern will shudder and change. By monitoring the movement in the interference pattern, it should be possible to detect the microscopic motions induced by gravitational waves. And it should be possible to do so with far more precision and accuracy than with Weber's bars.

Laser interferometry involved a completely new way of doing science, at least for relativists. Relativity had been a pencil-and-paper operation, with experiments few and far between. There were some laboratory setups and a few sparse collaborations between universities and institutions. It wasn't like particle and nuclear physics with the huge accelerators and reactors. But now, a new culture was necessary, one that could support spending tens or even hundreds of millions of dollars to construct experiments. Instead of teams with a handful of people, large organizations with hundreds of scientists and technicians would be required.

This time, it had to be done properly. This time, they had to know what they were looking for. It was clear that the gravitational waves had to come from something that pushed the theory to its limits. Hulse and Taylor's millisecond pulsars appeared quite benign, just two very compact stars orbiting each other. Yet they seemed to be able to spew out waves, enough of them to visibly suck out energy from their orbits. Neutron stars were stars almost on the brink of implosion that warped space and time enough to bring out the full glory of Einstein's theory.

One possible source of copious gravitational waves might be a supernova. Supernovae are imploding stars that for a few seconds shine more brightly than the billions of stars in our galaxy put together before becoming neutron stars or black holes. At any given time, a supernova is the brightest thing in the sky. Just as supernovae are a strong source of electromagnetic waves, astrophysicists speculated that they might be energetic enough to gnarl and shake spacetime into action, sending out a burst of gravitational waves. In 1987, a supernova went off in the nearby Large Magellanic Cloud, about 160,000 light-years away, and was observed in its full glory through normal telescopes. To everyone's embarrassment, not a single bar or other form of detector was running at the time to attempt to pick up the gravitational waves, except for Joe Weber's. Unsurprisingly, he claimed to have seen something, and as had become the custom, he was ignored.

The problem with supernovae is that they are too unpredictable, and while these huge explosions might indeed send a burst of energy, by the time a supernova's gravitational waves reached a detector on Earth they would be a mere blip. They could be confused with any other spurious bit of noise that might make its way to the instrument. No, what was needed was a clean signal that, even though it might be faint, would have a definite, perfectly known shape and form, like looking for a familiar face in a crowd.

There was something out there that might just do the job. The gravitational wave signal from the orbiting neutron stars Hulse and Taylor had observed could, in principle, be calculated with enough accuracy to be of use. Unlike the mess of waves coming out of a cosmic explosion, the gravitational wave signal should be regular and periodic, like a siren, and it would slowly change with time as the neutron stars lost energy and approached each other. The signal was simple, easy to describe, maybe even easy to detect.

But why stop there? Why not go for the big prize? A neutron star orbiting and plunging into a black hole would give a far stronger signal, and, of course, a binary system made up of two black holes would bring out the warpedness of Einstein's space and time in all its glory. Two black holes orbiting each other would send out a regular hum of gravitational waves. As they got closer and closer to each other, the pitch of the hum would get higher and higher until, just as they would be about to merge, they would send out a chirp and then a burst of waves that would evanesce as the black holes collapsed into one. This waveform is what the instruments would look for: the inspiral, the chirp, and then the ringdown. These relativistic binaries were like gems buried in the firmament. And the gravitational wave detectors would find them.

While it seemed straightforward — just look for the inspiraling neutron stars and black holes — a crucial piece of information was missing. What would the gravitational wave detector actually *see*? How exactly would the inspiral, chirp, and ringdown look once they arrived at the instrument? The observers, the new breed of gravitational wave astronomers, would need to know what kind of signal to expect, not roughly but exactly, if they were to be able to pull it out of the mess of noise that invariably polluted the data. And to be able to get a precise, exact answer to these questions it would be necessary to go back to the age-old problem of solving Einstein's field equations, this time to find precise mathematical solutions to describe what the gravitational waves would look like. Decades of experience showed that Einstein's equations seemed to turn and bite whoever tried to tame them. The only way forward was to solve the equations on a powerful computer and see what would happen when two black holes circled each other and ultimately collided.

Charles Misner, one of John Wheeler's students and collaborators, had already warned of the equations' treachery at the Chapel Hill meeting in 1957. You had to be careful trying to solve the gnarled, nonlinear beasts that Einstein had bequeathed, for, as Misner put it, there were only two possible outcomes: "either the programmer will shoot himself, or the machine will blow up." And the latter is exactly what had happened. In 1964, when one of Wheeler's ex-students, Robert Lindquist, tried to run the model, the program blew up. As the black holes got closer and closer to each other, the errors in the solutions became larger, and very quickly the computer was spewing out garbage — numerical diarrhea. The errors were so intractable that Lindquist gave up.

In the 1970s it was Bryce DeWitt's turn to try to find what would happen when two black holes collided on a computer. While his passion had always been for quantum gravity, he had learned how to simulate complicated equations on the computer during his work on the bomb project with Edward Teller at Lawrence Livermore National Laboratory in California. At Texas, he set one of his students, Larry Smarr, the task of working out how much gravitational radiation would be emitted if two black holes collided. They ran their code on the big University of Texas computer and were able to get a rough guess of what the gravitational waves would look like. And then the errors would blow up and garbage would come out. It was a glimpse of the waveform, but too rough to be useful. The singularities of spacetime would rear their ugly heads and kill the result.

For the next three decades, teams of programmers would work on trying to simulate the binaries and fail to get it right. Their work was advancing, but as Frans Pretorius, a relativist based at Princeton, recalls, "Naive things weren't working, no one exactly knew why, and people were sort of flailing about in the dark. And what made the problem so insidious was the computational expense of the full problem." In the 1990s the black hole collision problem was even considered one of the grand challenges of computational physics in the United States, with millions of dollars given to groups all across the country to buy supercomputers and run their programs. Every now and then there would be an improvement, and the results could progress a little bit further before the errors crept in. It became a field in its own right — numerical relativity.

Solving the equations for colliding black holes was difficult, unforgiving, as hard as detecting gravitational waves themselves and emblematic of Einstein's field equations. Young relativists would be sucked into trying to solve Einstein's equations on the computer and would spend their — often short — careers getting a small improvement on what had been done before. It was like playing an incredibly elaborate computer game, often on one's own, with no intermediate rewards, no passed levels, and no epic wins.

For some, general relativity came to mean numerical relativity. A general relativity group would not be complete without one or more relativists trying to solve the problem of black hole collisions on the computer with an eye on gravitational waves. There were conferences and meetings on the problem where everyone could get together to show off their new tricks and their plots and graphs. But the equations wouldn't give in. And with the waveforms that would come out of their simulations of binaries, there wasn't a hope in hell of finding them with the detectors.

Looking back at those dark times, Pretorius says, "There was a serious possibility that this was sufficiently difficult that it wouldn't be solved to a degree by the time [the gravitational wave detector] came online." The data could very easily precede a useful prediction of what the computer simulations might reveal.

Yet there was another side to the battle for numerical relativity that would have a surprising impact on the wider world. Through the late 1970s and early 1980s, Larry Smarr developed ever more elaborate numerical codes that he would attempt to run on the largest computers to which he could gain access. Based in the United States, Smarr found that he was doing many of his numerical runs in Germany, and his frustration grew at being unable to run his computer codes back home. By the mid-1980s, Smarr had successfully convinced the US government to fund a network of supercomputer centers to service all branches of science in need of "data crunching." Smarr would end up directing one of these new centers, the National Center for Supercomputing Applications in Illinois, and it was his research group that in the 1990s came up with the first graphical Web browser, Mosaic, that allowed them to visualize remote data over the Internet. And so, in the midst of the battle to conquer black holes, it was numerical relativity that gave birth to the Web-browsing culture that is such an integral part of our lives today.

While the numerical relativists flailed around, the plan to build an effective gravitational wave instrument was under way. This time, there could be no false discoveries exceeding the instrument's capabilities — the era of Weber was past. The interferometer was the method of choice, but the requirements for such a device were extreme. The laser light would have to travel far enough that a tiny deflection of the masses due to gravitational waves would be detectable in the interference pattern. Even with an interferometer that was kilometers long, the laser light would have to bounce back and forth, reflecting off mirrors tied to the masses, over a hundred times. The mirrors had to be perfect and perfectly aligned. And still the deflection would be tiny. A burst of gravitational waves coming from an inspiraling binary would lead to a deflection a minute fraction of the width of a proton.

A fully functioning interferometer that could truly detect gravitational waves coming in from outer space was an almost impossible beast to build. The laser beam would have to travel for kilometers at a time without deviating from its path by more than the width of an atom. The equipment would have to be set up as if it were floating on air, protected from all the ambient noises of everyday life, with perfect mirrors and state-of-the-art signal processing to be able to tease out the imperceptible deflections. It would have to be able to separate out the effect of the Earth's tides which could shift things around by a fraction of a millimeter, the rumbling of trucks on distant highways, and the vibrations from the electrical grid.

It would have to be perfect in every way, and it would have to be big. As interferometers began slowly to take over the field of gravitational waves, it became clear that their size and expense would limit the number that could be built. In Europe, the British and the Germans joined forces to build an interferometer with arm lengths of about 600 meters. Based near Sarstedt in Germany, it was named GEO600. A far bigger one named after the Virgo cluster of over a thousand galaxies, with arms of 3 kilometers, was conceived by the French and Italians and built in Cascina, Italy. In Japan, a smaller interferometer, TAMA, was built with arm lengths of 300 meters.

The poster child for gravitational wave interferometry was to be LIGO, the Laser Interferometer Gravitational Wave Observatory. It was originally led by two experimentalists, Rainer Weiss from MIT and Ronald Drever from Caltech, and the theorist Kip Thorne. First conceived in the early 1970s, LIGO had a difficult, fractious birth.

It was to be, by far, the grandest of all the interferometers. In fact, it wasn't one but two interferometers, one based in Hanford, Washington, and the other in Livingston, Louisiana. With two detectors located far apart, it would be possible to rule out results due to local noise, earthquakes, or traffic. And if it joined forces with one of the other detectors, like GEO600, it just might be able to pinpoint the direction of the gravitational wave sources and so would be a true observatory, a proper telescope. No one was yet sure *exactly* what they should expect to detect or whether the instrument would be sensitive enough. LIGO would have to be built in two steps. First they'd need to build a "proof of concept," a gigantic prototype that would work the way the relativists and experimenters wanted it to, a process that was expected to take more than a decade. Only after that could LIGO be upgraded and start looking for the interesting stuff. The projects would take a long time, but the payoff if LIGO actually saw gravitational waves would be staggering. Their detection would allow us to observe the universe in a completely new way, not using light or radio waves or any of the other conventional approaches. It would also be a completely new window on Einstein's general theory of relativity for, although most people believed gravitational waves were out there, no one had actually seen them directly. LIGO's discovery of gravity waves would be on a par with the discovery of the electron, proton, and neutron at the beginning of the twentieth century. It would be a Nobel Prize–winning experiment for sure.

The excitement over LIGO was not universal. The project was expected to cost hundreds of millions of dollars to build and run, draining funding from other research projects. LIGO inevitably took money away from the other gravitational wave experiments, but its impact on funding would also encroach on other fields. And by calling itself an observatory, LIGO was also stepping on the astronomers' toes. They could see LIGO sucking away precious cash from their own research. In a 1991 article in the *New York Times,* Tony Tyson from Bell Labs, who had worked on gravitational waves in the early days, wrote, "Most of the astrophysical community seems to feel it would be very difficult to get any important information from a gravitational-wave signal even if one should be detected." As Jeremiah Ostriker, a leading Princeton astrophysicist, said to the *New York Times,* the world "should wait for someone to come up with a cheaper more reliable approach to gravity waves." The astrophysicists were vocal, almost rabid in their opposition to LIGO. When asked to rank what astronomical projects should be given priority by the US funding agencies in the beginning of the 1990s, a panel of astronomers led by John Bahcall of the Institute for Advanced Study at Princeton didn't even bother to include LIGO in their rankings.

The American National Science Foundation turned down the first two proposals for LIGO and took five years from when the first proposal was submitted to finally approve a third proposal with a budget of $250 million, a seemingly exorbitant amount of money for an instrument that would quite probably see nothing and was, on the face of it, technologically impossible to build. Yet finally, in 1992, after almost twenty years of scheming, designing, and dreaming, the perfect experiment could go ahead.

Kip Thorne and his collaborators were already discussing their plans for LIGO when Frans Pretorius was born in South Africa. Pretorius grew up in the United States and Canada and completed his PhD at the University of British Columbia in Vancouver, learning the trade at one of the nerve centers of numerical relativity. He was offered a fellowship at Caltech, Kip Thorne's stomping ground, that let him do whatever he wanted. Pretorius decided to tackle the problem of inspiraling black holes on his own terms. In contrast to the big teams of computer programmers, working on the insurmountable problem of simulating the inspiral, chirp, and ringdown, Pretorius worked alone, "under the radar" as he recalls, not taking part in any of the big collaborations that were designing computer programs to solve the problem. Pretorius stepped back and looked at all the failed attempts of the past decades and picked out bits of different ideas that could be promising. He then set about writing a numerical program from scratch, in his own way, incorporating all of these ideas. He had an incredible instinct for what might and might not work. In his resulting code, Einstein's equations became much simpler, so simple that they looked almost like those of electromagnetism. And electromagnetic waves were easy to solve and evolve.

Then he ran it. It took several months for the program to run, a period Pretorius recalls as "pure agony." But to his growing surprise and elation, Pretorius was able to run his program all the way through, from the moment the black holes started inspiraling until they co-alesced, sent out a burst of waves, and then settled down into one fast-spinning black hole. There was the precise, accurate description of the gravitational waves that everyone had been so desperately looking for. Pretorius had finally solved Einstein's field equations on a computer. He had built on a battery of ideas that emerged before him, but it had taken his new, fresh look at the problem to put them together in exactly the right way.

Pretorius announced his results at a conference on general relativity in Banff, Alberta, in January 2005. Einstein's field equations had finally been cracked open, and it was possible for the first time to simulate two black holes orbiting one another, each sucking the other into its inexorable pull until the two coalesced into one, spitting out a barrage of gravitational waves that would gradually disappear with time. "There was quite a bit of excitement," Pretorius recalls. "People were interested enough to go outside of the talk to organize a session where people could ask all the detailed questions." Half a year later, two other groups announced that they had also been able to crack the problem using completely different methods of evolving the black hole binaries. Just like Pretorius, they were able to follow the catastrophic collapse of a pair of black holes all the way through. It was as if Pretorius's discovery had mentally unblocked all the work being done by other teams, and the results started to pour in, confirming Pretorius's calculation.

There was now a palpable sense of euphoria and relief. Finally, finally, it would be possible to describe the elusive waveforms. The observers would now know how to pick out the ghostly signals buried in the mayhem of noise measured by the interferometers.

Toward the end of his life, Joseph Weber came across as a bitter man. He bristled with anger at any discussion of gravitational waves. At the few conferences or workshops he attended, the audience would be subjected to decades of pent-up fury. He would rage at the mildest attempt to question him. He had seen gravitational radiation before everyone else and no one would take that away from him. Freeman Dyson, one of his early supporters, had in Weber's later life written to him pleading that he back down. Dyson had written, "A great man is not afraid to admit publicly that he has made a mistake and has changed his mind. I know you are a man of integrity. You are strong enough to admit that you are wrong. If you do this, your enemies will rejoice but your friends will rejoice even more. You will save yourself as a scientist."

Weber did no such thing. On the contrary, he had become the drag anchor of gravitational wave research, actively campaigning against LIGO. Weber had been in the press enough to have made a name for himself in the wider world as the expert on gravitational waves. When he spoke out, the powers that be would sometimes listen. In the early 1990s, when LIGO was making its third, desperate bid for funding, Weber wrote to Congress, stating that funding such a hugely expensive instrument would be a waste of money. His bars, he claimed, had seen gravitational waves and cost a fraction of a million dollars. There was no need to spend hundreds of millions. His ranting had little impact; throughout his career, Weber had made so many ludicrous claims that, as Bernard Schutz recalls, "by the time he was opposing LIGO, no one really wanted him on their side." If Weber felt ignored, he was making things worse for himself. He was now the enemy of the field he had created.

Weber died in 2000, before LIGO started operations. It had taken decades of devotion to get the most perfectly tuned instrument to work. Along the way, there had been delay after delay. Kip Thorne had made a number of bets with colleagues in the 1980s and 1990s that gravitational waves would be discovered before the turn of the millennium, and he lost them all. Even in the beginning of the twenty-first century, LIGO faced setbacks, from the loggers with their circular saws in the Louisiana forest who set off the detectors at Livingston, to mysterious whirrings in the nuclear reactors around the Hanford site in Washington. But when it was finally turned on in 2002 and run for a few years, LIGO was able to achieve the sensitivity everyone had been gunning for. It was the first stage in the experimental journey laid out in the proposal in the early 1990s. Its detectors could pick up vibrations of less than a proton's width, as had been envisioned decades before. In fact, the LIGO team announced, the instrument was even more sensitive than they had predicted. LIGO was, by all means, a resounding success, even though it didn't see anything. As expected in its first incarnation, LIGO was not yet sensitive enough to actually detect gravitational waves, but it did show the way forward. The LIGO team can now improve the existing instrument so that at some point it will see the ripples in spacetime that Einstein had first predicted.

It is a long game. Unlike Weber's results, which came fast and steady the moment he turned on his instrument, LIGO will have used up thousands of technicians over many decades before it can actually detect gravitational waves. The founding trio, Ron Drever, Kip Thorne, and Rainer Weiss, now in their seventies and eighties, might not all be around when that moment comes, and they may have devoted their lives to something they will never see. But there is unwavering confidence that waves are out there; Einstein's theory predicts them, and they have been seen, albeit indirectly, through the gentle but steady orbital decay of the millisecond pulsars. It is just a matter of time before gravitational waves are seen, and then a field of research that started with Weber's bang will end with a whimper: the whimper of spacetime shimmering as it passes through Earth.

grand picture of the universe. Cosmology's task is difficult, controversial, and unfinished, but, as Rees argued, it is also of the utmost importance.

The picture of the universe that cosmology was revealing by the time of the Princeton conference was truly bizarre. It seemed that we understood far less of the universe than we had originally thought. In fact, a large fraction of the universe appeared to be in the form of exotic substances we had never seen in a laboratory. Dubbed "dark matter" and "dark energy," they were out there, affecting spacetime, yet strangely elusive and undetectable. The case for a dark universe emerged forcefully one afternoon when the large-scale structure of the universe was discussed. It was that one topic that had drawn me into cosmology in the first place.

When we look out at the universe, we see an elaborate tapestry of light, with galaxies clumped into clusters, filaments, and walls, leaving large voids of emptiness. It is rich, full of information and complexity. Where does this large-scale structure of the universe come from? This was the most pressing question for the conference attendees, for the answer was still completely up for grabs, and the conference organizers dedicated a full afternoon to the topic. J. Richard Gott, a tall, gangly astronomer from Princeton with a deep and slow southern drawl, stood up and defended common sense. At a first glance the universe looks very empty, so Gott proposed a universe almost completely devoid of matter that slowly evolved to form a tapestry of galaxies and clusters of galaxies that would populate the night sky. Another young and energetic astronomer from Princeton named David Spergel proposed that the universe is not at all empty, but rather full of an invisible, dark form of matter. Spergel's dark matter would be made up of some fundamental particle unaccounted for in the standard model of particle physics that had not yet been observed in any experiment. But it was the final speaker, Michael Turner, a sharp-witted theoretical cosmologist from Chicago, who made the most outlandish proposal of the afternoon: Why not assume that the universe is permeated by the energy of a cosmological constant? In Turner's universe, about two-thirds of the overall energy would be accounted for by the constant

Einstein had so firmly rejected almost seventy years before. The crowd was not impressed with Turner's proposal. *Anything but a cosmological constant* — it was Einstein's biggest blunder.

Chairing the gladiatorial combat between the universes was Phillip James (Jim) Peebles, then the Albert Einstein Professor of Science at Princeton University. A tall, slim man with a thoughtful face lifted from a portrait by Modigliani, Peebles was the consummate gentleman, courteously moderating the debate. While he was careful to the keep the conversation on track, he would sometimes chuckle with almost childish glee at the jibes and comments being thrown across the stage. The Critical Dialogues meeting was partly organized to celebrate Peebles's sixtieth birthday, a fitting tribute. For the previous three decades, Peebles had been the prime architect of the theory of large-scale structure of the universe at the heart of modern cosmology.

In the early 1970s, Jim Peebles published a slim volume, *Physical Cosmology*, a summary of a set of graduate lectures he gave at Princeton in 1969. John Wheeler had attended, taken notes, and, according to Peebles, bullied him into publishing the lectures. In the introduction to *Physical Cosmology*, Peebles briefly mentioned the cosmological constant, saying that "the cosmological constant Λ [the Greek capital letter "lambda," which is the mathematical symbol for the cosmological constant] is seldom mentioned in these notes." For Peebles, the constant was an unnecessary complication, "the dirty little secret" of cosmology. Everyone knew that the mathematics allowed for it, but because it made the physics too bizarre and troublesome, everyone pretended it wasn't there. Now, a quarter of a century later, despite being reviled by the majority of Peebles's colleagues, the cosmological constant was about to make a comeback. It would do so with a vengeance.

When Jim Peebles arrived in Princeton in 1958, fresh out of engineering school at the University of Manitoba, he found John Wheeler and his crew chipping away at black holes and the final state. Wheeler was not the only acolyte of general relativity at Princeton; there was also Robert Dicke. Like Wheeler, in the mid-1950s, Dicke realized what dire straits Einstein's theory was in, with little or no progress being made in testing it. He created his own gravity group at Princeton, where general relativity could be discussed and, most important, measured and tested. "Rather quickly in my career I got into orbit around Bob and into doing things that were exciting," Peebles says. He joined Dicke's team as a PhD student and, after graduating, focused his research on testing gravity physics. He would stay in Princeton for the next fifty years.

In the 1960s, Peebles recalls, cosmology was still "a limited subject — a subject, as it used to be advertised, with two or three numbers," and, Peebles says, "A science with two or three numbers always seemed to me to be pretty dismal." There were few people actively working in the field, and very little research was under way. This suited Peebles just fine. He could devote himself privately and quietly to tackling problems that took his fancy at his own pace. Having completed his PhD on quantum physics, from then on Peebles devoted himself to fleshing out cosmology. He started with what his colleagues at Princeton called the "primeval fireball," working out what actually happened to atoms and nuclei in the very early universe when it was hot and dense. He worked like a craftsman. Shut away in his office, he filled page after page with handwritten equations, slowly going over his calculations and honing his approach.

Peebles's mentor took a different approach. As Peebles recalls, "To him physics was certainly theory but it had to lead to an experiment that could be done in the near future," so Dicke had his team look for the relic radiation left over from the primeval fireball. They developed a new form of detector that could scan the sky from the roof of the physics building, but they didn't find the radiation in time. One Tuesday in late 1964, Dicke's team was sitting in his office for their weekly meeting when the phone rang. Dicke picked up the phone and spoke to someone for a few minutes. "We've been scooped," he said when he put the phone down. Arno Penzias had just called to tell him that, with Robert Wilson at Bell Labs, he may have just found evidence for the relic radiation. Within months Dicke and his team had confirmed the result from Bell Labs, but it was too late: Penzias and Wilson would go on to win the Nobel Prize on their own.

To Peebles, there was something wrong with the picture of the cosmos that appeared in 1960s physics textbooks. At the time, there were two completely different topics. On the one hand there was the history and evolution of the universe, the story that Friedmann and Lemaître had told. It explained how space, time, and matter evolved on the largest possible scales. On the other hand there was the stuff the astronomers looked at, galaxies and clusters of galaxies. While these galaxies are part of the universe, their presence seemed almost superficial and unconnected to the fundamental development and structure of the universe, like rich, colorful swirls of light painted on spacetime. It was true that galaxies told us a lot about the universe, such as how fast the universe was expanding and how much stuff it actually contained. But, looking up at the sky, Peebles felt that there had to be more to galaxies — he was convinced they must play a key role in the evolution and large-scale structure of the universe, and surely their own origin must be connected to it as well. They couldn't have appeared out of nothing, great blobs of light, gas, and stars dropped into spacetime as an afterthought. This meant that galaxies must also play a role in Einstein's general theory of relativity. The question was how. This was a perfect challenge for Peebles: a difficult, open problem that hardly anyone wanted to work on.

The role of gravity in individual galaxy formation is obvious. A collection of matter collapses under the pull of its own gravity. If there's enough matter, and it has enough kinetic energy to avoid collapsing below a certain point, the resulting blob becomes a galaxy, reined in by its own gravitational pull. What was less clear when Peebles approached the topic was how the gravitational effects in individual galaxy formation related to gravity's role in the expansion of the universe as a whole. The Abbé Lemaître had pointed out that there must be a connection, and the Russian theorist George Gamow had mused on how galaxies would form in an expanding universe, but neither could provide a proper calculation to back up their speculations. In 1946, Evgeny Lifshitz, one of Lev Landau's disciples, had taken Einstein's field equations and attempted to link what happened on the scale of the universe with the much smaller scale of individual galaxies. His result hinted at how the large-scale structure of the universe would emerge — small ripples in spacetime would evolve and grow, following his equations, and galaxies would end up forming and clustering in regions of high curvature to create the large structures that can be observed today.

When Peebles worked out how atoms and light would have behaved in the early universe, he realized this new understanding of the hot early universe might explain how galaxies formed shortly after the Big Bang. When Peebles put in some rough estimates for the age of the universe, the density of atoms, and the temperature of the relic radiation, he found that collapsed structures *could* form with masses between a billion and hundreds of thousands of billions times that of the sun, just like the Milky Way. As Gamow had previously surmised, the early universe appeared to be an ideal breeding ground for galaxies.

As Peebles continued to figure out the details of how galaxies formed, he was not alone. A young PhD student at Harvard named Joseph Silk argued that the collapsing blobs that would ultimately form galaxies should also leave an imprint on the primeval fireball — a faint patchwork of hot and cold regions in the relic radiation that had recently been discovered by Penzias and Wilson. Silk's results were echoed by Rainer Sachs and his student Arthur Wolfe at Austin, who found that even on the largest scales, the relic radiation would be affected by the gravitational collapse of all the matter in the universe. Yakov Zel'dovich's team in the Soviet Union was also finding the same thing. Their results indicated that by looking at the ripples in the relic radiation left over from when the universe was a few hundred thousand years old, it would be possible to see the first moments that led to the formation of galaxies. In a scattered and disjointed way, Gamow and Peebles's physical cosmology was beginning to bear fruit.

Peebles wanted to explain the expansion of the universe — the hot beginning, the primeval fireball, the atoms, the gravitational collapse — in terms of basic textbook physics, combining general relativity, thermodynamics, and the laws of light. With a PhD student from Hong Kong named Jer Yu, Peebles wrote out the complete set of equations that would allow him to work through the evolution of the universe from the earliest moments after the Big Bang until today. Peebles's universe starts off in a smooth, hot state with a very small set of ripples disturbing the primordial slush of gas and light. As these disturbances evolve, they encounter pressure from the messy, sticky plasma of free electrons and protons. The universe vibrates with waves like a rippling pond until the moment electrons and protons combine to form hydrogen and helium. Then the next stage begins: atoms and molecules start to clump together, collapsing under the pull of gravity, creating nuggets of mass and light scattered throughout spacetime. These are the galaxies and clusters of galaxies that emerged from the hot Big Bang.

In Peebles and Yu's universe, the way that galaxies are scattered in space to form the large-scale structure of the universe should carry with it the memory of the universe's hot beginning. The relic radiation left over from the Big Bang, which Penzias and Wilson had measured to have a temperature of just 3 degrees Kelvin, should carry an echo of the small ripples that seeded the formation of galaxies. By solving the equations of the universe in one consistent, coherent whole, Peebles and Yu found a new, powerful way of studying Einstein's theory of general relativity: look at how galaxies are distributed in space to form the large-scale structure of the universe and use it to discover how spacetime began and evolved.

It was a powerful, compelling narrative, but Peebles and Yu's results were met with silence. "No one paid any attention to our paper," recalls Peebles. In bringing together the different areas of physics, Peebles and Yu had wandered into an intellectual no man's land. Their work wasn't strictly astronomy, nor was it general relativity or fundamental physics. The lack of response was fine by Peebles. He continued working on the universe, occasionally roping in the odd student or young collaborator, but for the most part quietly and peacefully calculating away on his own.

Now that Peebles had a model of the universe, he needed to look at some data to see if he was on the right track. In the early 1950s, the French astronomer Gérard de Vaucouleurs, based at the University of Texas, had looked at a particular catalogue of over a thousand galaxies, the Shapley-Ames Catalogue, and found a "stream of galaxies" stretching across the sky, bigger than any cluster, more like a "supercluster" or "supergalaxy." His work was not well received. Walter Baade, a Caltech astronomer, dismissed the result, saying, "We have no evidence for the existence of a Super Galaxy," as did Fritz Zwicky, who simply asserted, "Superclustering is nonexistent." Peebles was skeptical about de Vaucouleurs's result, but as one of his students recalls, Peebles would echo the view of his mentor, Bob Dicke, that "good observations are worth more than another mediocre theory." So he set out on a quest to map out the large-scale structure himself, with his protégés, sometimes with surprising results. When Marc Davis and John Huchra, both young researchers at Harvard, found that indeed there were immense structures in the far crisper surveys of galaxies they were producing, Peebles was "flabbergasted." As he acknowledged, "I wrote some pretty vitriolic papers with examples in the past of how astronomers had been misled by just this tendency . . . to pick patterns out of noise. It was clear you needed a pattern forming mechanism." But with time, he realized that galaxies were indeed arranged in a vast tapestry of walls, filaments, and clusters, what became known as the cosmic web. The large-scale structure that Peebles had predicted in his computer models was beginning to emerge in the real world.

In 1979, Stephen Hawking, along with a South African relativist named Werner Israel, put together a survey of relativity to celebrate Einstein's centenary. They brought together the leading researchers in cosmology, black holes, and quantum gravity. Bob Dicke and Jim Peebles contributed an essay titled "The Big Bang Cosmology — Enigmas and Nostrums." It was a short essay. In a few pages, Dicke and Peebles laid out what they believed to be some fundamental problems in an incredibly successful theory.

So what was wrong? For a start, the universe seems far too smooth. Although there had been attempts to come up with an explanation in the past, Dicke and Peebles couldn't to their satisfaction identify one that worked. And there was more. Why does the geometry of space, as opposed to that of spacetime, look so simple? The geometry of space seems to have no overall curvature, and the rules of high-school-level Euclidean geometry apply. Such rules as *Parallel lines never intersect* and *The sum of the angles of a triangle is 180 degrees* seem invariably true. A universe with no spatial curvature is allowed in general relativity, but it is a very special case. Einstein's equations predict that the evolution of the universe is likely to push the curvature away from zero incredibly quickly. So, if the universe seems to have almost no curvature today, it must have had even *less* curvature in the past. The universe we live in is extremely unlikely. Finally, the galaxies and structures built up of galaxies spanning the heavens must have come from somewhere. Conditions had to be perfectly tuned for the universe to look as it does today. At the Big Bang, the tendency of the universe to expand had to be just enough to compensate for the pull of gravity and prevent the whole of spacetime from collapsing into itself, yet not so extreme that spacetime would fly apart in an empty void. Their article boiled down to a simple question: What happened in the very beginning?

Dicke and Peebles's article was followed by another short essay by Yakov Zel'dovich. In his article, Zel'dovich pondered the very early universe following the line of reasoning that the Abbé Lemaître had first taken when discussing his primordial atom. There was a whole plethora of interesting phenomena at play in the hot early universe that could impact its evolution and affect how it evolved into what we see today. Zel'dovich urged the community of particle physicists and relativists to figure out what these effects would be.

Dicke and Peebles's and Zel'dovich's papers were prescient. Just one year later, cosmology would be turned on its head by a simple proposal for how the early universe evolved. The idea had been floating around in an unformed way, but it took Alan Guth, a postdoc at the Stanford Linear Accelerator Center, to come up with the essence of cosmic inflation. Guth realized that in some grand unified theories — theories that attempted to unify the electromagnetic, weak, and strong forces into one overarching force — the universe could be trapped in a state in which the energy of one of the fields was incredibly high and dominated everything else. In that state, the universe would be driven to expand rapidly, or inflate, as Guth dubbed it. Although Guth's original idea turned out to be flawed — if the universe was trapped in such a state, there was no way of getting out of it — new ways of making the universe inflate were quickly proposed by others.

The idea of an inflating universe, or inflation, opened up a new avenue in cosmology, revealing a new period in the universe's past that could be explored. Now there was a theory that predicted exactly how the universe should be when structure started to form, and it seemed to address the problems raised by Dicke and Peebles. For a start, the theory of inflation pushed space to almost instantaneously have no curvature. Imagine taking a round balloon that you can hold in your hands and using a giant pump to blow it up so quickly that it almost instantly becomes the size of the Earth. From your perspective, the piece of balloon in front of you would now look pretty flat. Inflation would also drive the universe toward a tremendously smooth, pristine state. Any large lumps or voids that would naturally pepper the land- scape of spacetime would have been pushed far out into the distance, invisible to our gaze. Inflation also brought with it a way to kick-start the growth of structure in the very early universe. During the period of intense inflation, the microscopic quantum fluctuations in the fabric of spacetime would be stretched and imprinted onto the largest scales.

Inflation, as astrophysicists in Chicago succinctly put it, established the link between "inner space and outer space." Inner space was the world of the quantum and the fundamental forces, and outer space encompassed the cosmos, where general relativity came into its own. And so, the program of research that Peebles had been developing over the previous decade, along with the work of Zel'dovich, Silk, and oth- ers, took on a new purpose: the large-scale structure of the universe, the distribution of galaxies, and relic light should hold the clues that link inner and outer space. People began to take notice.

In 1982, Peebles tried to construct a new universe. The old model he'd developed with Jer Yu, made of atoms and radiation, wasn't working out. When he compared the results of his model to the surveys of gal- axies that had been mapped out in the sky, they didn't match. Reality simply didn't agree with his elegant calculation. Not only that, in the previous decade, galaxies themselves seemed to have become a whole lot more complicated. A strange picture was emerging of what was going on inside them.

The American astronomer Vera Rubin had found that galaxies seemed to spin far too quickly for their own good, like manic Cath- erine wheels held together by a mysterious force. Rubin focused her telescope on the Andromeda Galaxy, a swirl of stars and gas spinning at hundreds of kilometers per second. At least that's how it appears if you look at it with a telescope. There was much more light at the center where all the stars are concentrated, so Rubin expected that most of the gravitational pull keeping the galaxy together would come from its central core. But as she looked at nuggets of stars farther and farther away from the center of the galaxy, she found they were moving far too quickly. In fact, the stars were speeding around so quickly that Rubin simply couldn't understand how the gravitational pull of the galaxy's center could rein them in. It was as if the Earth suddenly doubled or tripled the speed of its orbit around the sun. Unless the sun somehow increased its gravitational pull, the Earth would simply fly out of the sun's orbit and shoot off into space. Something else, big and invisible, was holding the outer stars in their orbits.

Fritz Zwicky observed a similar phenomenon in the 1930s, but his results were ignored for almost forty years. Zwicky had looked at the Coma cluster of galaxies and added up the total amount of mass he could see there. He had then measured the speed with which the galaxies were moving around inside the cluster and found that they were moving far too quickly. As he said in a paper he published in Switzerland in 1937, "The density of luminous matter in Coma must be minuscule compared with the density in some sort of dark matter."

Jim Peebles was coming up against his own problems with galaxies. With a young collaborator from Princeton, Jerry Ostriker, he set about building simple computer models for how galaxies formed, representing them as a bunch of particles pulling each other through gravity and spinning around in a spiral. But whenever he set his models spinning, the galaxies would disintegrate. A blob would form at the center that would stretch out through the arms and tear the galaxy apart. Ostriker and Peebles tried to stabilize their models by immersing their spinning particles in a ball of invisible mass. This sphere of stuff—a halo, they called it—would bolster the gravity keeping the galaxy together. The halo had to be dark (that is, invisible) so as not to be detected by telescopes. Paradoxically, the model showed that this dark matter had to be much more abundant than the atoms that were seen in stars. In the late 1970s, Sandra Faber, working at Santa Cruz in California, and Jay Gallagher, working in Illinois, wrote a review in which they collated the odd findings that astronomers were getting when looking at galaxies and that Peebles and his colleagues were discovering when simulating them. They concluded that "we think it likely that the discovery of invisible matter will endure as one of the major conclusions of modern astronomy."

In 1982, when Peebles began building a new model of the universe, he decided to include atoms *and* dark matter. In fact, he assumed that almost *all* of the universe was made up of a mysterious form of matter composed of heavy particles, invisible to us because it didn't interact with light. Peebles's cold dark matter model was simple, and it enabled him to predict what the distribution of galaxies looked like and how large the ripples in the relic radiation would be. This approach would prove to have a momentous impact on the development of cosmology, but as Peebles recalled, "I didn't take it at all seriously . . . I wrote it down because it was simple and it could fit the observations."

While Peebles didn't refer to the recently proposed inflationary era, his new model fit the Zeitgeist perfectly. It invoked a massive particle that could arise from fundamental physics, connecting inner and outer space. The cold dark matter model, or CDM model for short, was adopted by a growing army of astronomers and physicists who began to work out the fine details of how galaxies would actually form. Marc Davis at Berkeley allied himself with two British astronomers, George Efstathiou and Simon White, and the Mexican astronomer Carlos Frenk to build computer models to follow the formation of individual galaxies and clusters of galaxies in virtual universes. In their simulations, this gang of four, as they became known, would follow hundreds of thousands of particles as they interacted with one another, coming together to form the large-scale structure of the universe.

While CDM was popular and eagerly adopted, too many things seemed to go wrong. In the CDM model Peebles created, the universe could be only 7 billion years old, which was far too young. Astronomers had found dense pockets of stars known as globular clusters bobbing around in galaxies. These bright concentrations of light were full of old stars that must have formed early on in the history of the universe when it was mostly full of hydrogen and helium, which meant that the globular clusters had to be at least 10 billion years old. And there was more. If the universe was primarily made up of cold dark matter, the proportion of dark matter to atoms would be roughly 25 to 1. Yet, hard as they looked, astronomers couldn't figure out where that dark matter was. From the speed at which galaxies rotated or from the temperature of clusters of galaxies they observed they could infer how much gravity there was (the hotter they were, the more gravitational pull there had to be) and how much dark matter was necessary to generate that amount of gravity. The ratio of dark matter to atoms they kept on coming up with was closer to 6 to 1. True, the methods for weighing the dark matter were still crude and uncertain, but the deficit seemed too great to be explained by the margin of error. Almost immediately after creating the CDM model, Peebles felt compelled to give it up and look for alternative models. "There was a lot of net casting in the eighties and early nineties," as he put it.

The gang of four didn't fare any better. They used their computer models to create virtual universes and compared them with the real universe to see if they looked alike. They didn't. For a start, the real universe appeared to be much more structured and complex on large scales than the fake universes. In the CDM universe, the galaxies were much more clustered on small scales but smoothed out more quickly, once you zoomed out to look at the bigger picture, than in the real universe. It was possible to alleviate some of the problems in the virtual universes by slightly fudging the results, but the truth was that Peebles's simple model wasn't entirely working.

Despite the fact that it conflicted with basic observations, the cold dark matter model was embraced by the majority of astronomers and physicists. It was conceptually simple and fit nicely with inflation and the evidence for dark matter in galaxies. CDM's adherents looked for ways to further develop and somehow fix the model. One way of fixing the CDM involved resurrecting Einstein's cosmological constant. To many, that was anathema.

The case against the cosmological constant had become stronger since Einstein first introduced it in 1917. While he had, with the discovery of the expanding universe, rapidly discarded the cosmological constant from his theory, a few of his colleagues clung to it. Both Eddington and the Abbé Lemaître chose to incorporate it in their models of the universe. Lemaître went so far as to conjecture that the cosmological constant was nothing more than the energy density of the vacuum. In 1967 Zel'dovich showed what a serious problem the cosmological constant could be. He added up the energy of all the virtual particles that would pop in and out of existence in the universe and found that the resulting energy density would look like a cosmological constant but should have a truly gigantic value. Strictly speaking, the resulting cosmological constant would be infinite, for exactly the same reasons that everything involving quantum gravity was infinite, but a little hand waving could make it finite. Even so, it was a huge number, orders of magnitude greater than any energy that had been measured in the cosmos.

Zel'dovich's calculation showed that if there was an energy of the vacuum in the universe — and therefore a cosmological constant — it would be far too big to be compatible with observations. The only way to proceed was to assume that some as-yet-undiscovered physical mechanism intervened to make the cosmological constant equal zero. In practice cosmologists chose to ignore the cosmological constant and pretend it didn't exist.

Yet, again and again, whenever anyone tried to resolve the problems with the cold dark matter model, the cosmological constant, known as lambda, always cropped up as one of the possible solutions. In 1984, Peebles himself found that a viable universe with cold dark matter would need lambda to make up about 80 percent of the total energy of the universe. When the gang of four — Davis, Efstathiou, Frenk, and White — tried simulating one of their universes with lambda in it, they found that many of the problems they came up against with the simple CDM scenario went away.

In 1990, George Efstathiou, then at the University of Oxford, published a paper in *Nature* called "The Cosmological Constant and Cold Dark Matter." In it, Efstathiou and his collaborators compared the large-scale structure from a simulated universe, including the cosmological constant, with the real universe, this time using a catalogue with millions of galaxies that they had collected over the years. In their opening salvo, they claimed, "We argue here that the successes of the CDM theory can be retained and the new observations accommodated in a spatially flat cosmology in which as much as 80% of the critical density is provided by a positive cosmological constant," and they proceeded to show that such a universe seemed to fit all the observational data then available. Jerry Ostriker and Paul Steinhardt, one of the fathers of the theory of inflation, published a paper in *Nature* in 1995 where they argued that "a universe having critical energy density and a large cosmological constant appears to be favoured." Everything seemed to point to lambda.

While hints of lambda were appearing in large-scale structure, everyone shied away. As Jim Peebles wrote in 1984, "The problem with the choice ... is that it does not seem plausible." As Efstathiou and his colleagues stated in the conclusion of their paper, "A non-zero cosmological constant would have profound implications for fundamental physics." In another paper, George Blumenthal, Avishai Dekel, and Joel Primack from Santa Cruz in California argued that having a cosmological constant "requires a seemingly implausible amount of fine-tuning of the parameters of the theory." Indeed, as Jerry Ostriker and Paul Steinhardt wrote, the observational evidence opened up an impossible challenge: "How can we explain the non-zero value of the cosmological constant from a theoretical point of view?" The problem couldn't be kept a dirty little secret anymore.

At the Princeton meeting in 1996, Michael Turner from the University of Chicago faced a barrage of abuse as he sparred with Richard Gott and David Spergel in defense of the cosmological constant. The observations were in his favor, but the cosmological constant remained too unpalatable for his fellow cosmologists. It was too conceptually impossible and too aesthetically unpleasing. He probably would have gotten off more easily if he had called for divine intervention. At the end of the debate, the standard, cosmological-constant-free CDM model was declared the victor. Jim Peebles watched the spectacle in fascination.

By 1996, cosmology had been transformed beyond Jim Peebles's wildest expectations. He had started off, along with Yakov Zel'dovich, Joe Silk, and a few others, as one of the lone pioneers building up the theory of large-scale structure. Peebles had effectively made up the techniques that were used not only to theorize but also to analyze observations. Now a new generation of theorists was pushing his ideas forward with alarming ferocity while the astronomers were mapping out the universe with ever-increasing precision.

In this new era, Peebles found himself in the odd position of a contrarian in a field he had helped create. He disliked the fervor with which the CDM model had been adopted by his colleagues and continuously put forth new models to compete with it. But, as his mentor, Bob Dicke, had said, good data would trump all. CDM's supporters and Peebles were both about to be trumped.

In 1992, George Smoot, one of the principal investigators on the *Cosmic Background Explorer,* or COBE for short, claimed, "If you're religious, this is like looking at God." COBE was a satellite experiment designed to detect the relic radiation left over from the Big Bang with unprecedented precision and to map how its brightness would change as you looked in different directions in the sky. What Smoot was talking about was the first-ever measurement of the elusive *ripples* in the relic radiation, the small imperfections that Peebles, Silk, Novikov, and Sunyaev had for the previous twenty-five years been saying should be out there. It had been a long and almost embarrassing search. As time passed and the ripples remained invisible, the theorists had reworked their predictions, downgrading their expectations. In 1992, the COBE satellite, using a set of detectors based on Bob Dicke's ideas, made a map of the relic radiation, and there was a collective sigh of relief. Smoot went on to win the Nobel Prize for his work on COBE.

COBE's discovery was just the beginning. The picture it provided of the ripples in the relic light was still blurred and unfocused. The ripples needed to be brought into focus, for, as Peebles, Novikov, and Zel'dovich had shown, there should be a rich tapestry of hot and cold spots in the relic light that could be used to chart out the geometry of space. If the geometry of space was truly Euclidean, the size of the spots should subtend an angle of about 1 degree on the sky. And measuring the geometry of space was tantamount, through general relativity, to measuring the amount of energy in the whole universe. Better experiments were needed. Dozens of groups throughout the world developed instruments that could measure the relic radiation with better precision and focus. It was as if a band of intrepid explorers had set out to chart a new continent that had just been discovered. When, at the turn of the millennium, it finally all came together, a clutch of experimenters announced the discovery that the hot and cold spots indeed had an angular size of about 1 degree, and therefore the geometry of space had to be flat. The result was just as inflation had predicted and further evidence from the large-scale structure of the universe for CDM *and* a cosmological constant.

The final piece of data that definitively tipped the balance in favor of the cosmological constant came not from the field of large-scale structure that Peebles had so lovingly built up but from exploding supernovae in the distant universe. The first hint came in January of 1998, at the annual meeting of the American Astronomical Society, when a West Coast–based team of astronomers and physicists called the Supernova Cosmology Project claimed that there wasn't enough gravitational pull from dark matter or atoms to rein in and slow down the expansion of the universe. In fact, the Supernova Cosmology Project was finding that expansion of the universe was quite possibly accelerating. This meant that the universe was either much emptier than previously thought or had a cosmological constant that was driving space apart.

The Supernova Cosmology Project was to some extent simply repeating what Hubble and Humason had done in the 1920s: measuring the distances and redshifts of distant objects. Instead of looking at galaxies, the observers now had to look for individual supernovae, stars that exploded with an intense burst of light as bright as a whole galaxy concentrated in a pinprick, and that could be seen at far greater distances than ever observed by Hubble and Humason. While, in spirit, the work of the Supernova Cosmology Project echoed that of Hubble and Humason, this was no longer a two-person job, but a large operation with teams spread out over three continents using many Earthbound telescopes as well as the Hubble Space Telescope to produce their numbers. The measurement methods were difficult and had taken over a decade to perfect.

The Supernova Cosmology Project was closely followed by the High-Z Supernova Search project, which was finding similar results:

tentative evidence for accelerated expansion of the universe and, therefore, a cosmological constant.

Neither team could bring themselves to announce what they saw in their data. At the AAS meeting in Washington, in January 2008, their presentations were cautious, almost painfully so. The true implication of their results was quietly discussed in the corridors and made its way into the newspapers. The day after the announcements by the supernova teams, the write-up in the *Washington Post* said, "The findings also appear to breathe fresh life into the theory that there is a so-called cosmological constant." A few weeks later, *Science* magazine went further, publishing an article with the title "Exploding Stars Point to a Universal Repulsive Force." In the article, the leader of the Supernova Cosmology Project, Saul Perlmutter, refused to go so far, simply commenting, "This needs more work."

Just over a month later, the High-Z team came clean and said it: there was lambda in their data. Not only was the universe too empty of atoms and dark matter, it was full of something else that was making it accelerate. Members of the High-Z team were invited on television around the globe to explain their strange, unfathomable results to the general public. CNN announced that scientists were "stunned the universe may be accelerating," and the leader of the High-Z, Brian Schmidt, was quoted in the *New York Times* as saying, "My own reaction is somewhere between amazement and horror. Amazement, because I just did not expect this result, and horror in knowing that it will likely be disbelieved by a majority of astronomers — who, like myself, are extremely skeptical of the unexpected." The SCP rapidly followed suit with its own results. It was official: lambda was out there. For their discovery the leaders of the two teams, Saul Perlmutter, Brian Schmidt, and Adam Riess, were awarded the Nobel Prize in 2011.

For years, even decades, there had been uncertainty about the universe's makeup, age, geometry, and basic constituents. All the different proposals had their pros and cons, and cosmology had become as much a matter of aesthetics as science, with practitioners choosing their preferred theories according to personal taste. But now the most unpalatable theory of them all, the cosmological constant, had won out. Within months, a new standard model of cosmology, known as the concordance model, or unimaginatively "Lambda CDM," had taken root. This new model of the universe contained a cocktail of atoms, cold dark matter, and a cosmological constant. It was the universe that large-scale structure had been hinting at for a decade but that hardly anyone had been ready to embrace. Even Peebles, with his unwillingness to follow the herd, was amazed at how everything had come together. But it was the data that had done it, exactly as his mentor had said it would. Peebles had to admit, "The best explanation for what the data is telling us is a cosmological constant. Or something that looks like a cosmological constant."

When Jim Peebles retired from teaching at Princeton, in 2000, he spent more of his time going on walks and taking pictures of wildlife. He relished the beauty and sometimes strangeness of the birds that he would stumble across on his treks, and now he had more time to do so. Instead of focusing on the patterns that galaxies traced in the sky or the ways that individual galaxies spun, he could lose himself in the surrounding beauty of woods and forests. It was this careful gaze and attention to detail that had helped him oversee the transformation of cosmology into a hard, precise science. Yet another strand of general relativity had matured and gained a life of its own. Peebles's quiet and persistent effort, his "scribbling," as he liked to call it, had placed the study of the large-scale structure of the universe firmly at the center of physics and astrophysics. The maverick in him had guided the field toward the bizarre model of the universe that had taken root: a universe in which 96 percent of its energy was in some dark substances, a combination of dark matter and the cosmological constant. Compared to when he had started off, almost fifty years before, it was a surreal turn of events.

The cosmological constant was now universally accepted. The fundamental problem remained: the gross inconsistency with what Zel'dovich had predicted from adding up the energy of the virtual particles in the universe and the value that was actually observed, a mismatch of over a hundred orders of magnitude. But while, in the past, this inconsistency had led cosmologists to not even consider the possibility of the cosmological constant, now they embraced it. It was there, in the data, unavoidable. In their textbook on relativistic astrophysics, written in 1967, Yakov Zel'dovich and Igor Novikov had said, "After a genie is let out of the bottle . . . legend has it that the genie can be chased back in only with great difficulty." There was truth in this analogy. Now, with the general shift toward the concordance model, the cosmological constant had to be tackled head-on.

Or maybe not. One more effort to yet again avoid the cosmological constant invoked an altogether new type of stuff that was pushing space apart. This exotic new field, particle, or substance behaved very much like a cosmological constant, but it was soon widely referred to as "dark energy." There were, and are, high hopes for dark energy and its potential to link the successes of observational cosmology with the creativity of particle physics and the quantum. Young and old cosmologists flocked to work on the topic in droves; in one talk at a conference, a speaker put up a slide with over one hundred different models for dark energy, a testament to the creativity of the new generation of cosmologists. And yet the invention of dark energy still didn't solve the problem that Zel'dovich had raised, that the energy of the vacuum was, in principle, far too big to be acceptable. Once again, the approach was to pretend the discrepancy wasn't there. It would take a revolution in the quantum theory of gravity to come up with a controversial solution.

The rise of physical cosmology in the past forty years transformed the way we look at spacetime and the universe. In mining general relativity on the grandest of scales and carefully teasing out the large-scale properties of the universe, Jim Peebles and his contemporaries opened up a completely new window on reality. Allied with the stupendous successes in mapping out the distribution of galaxies and the relic radiation, their work has revealed a bizarre universe, full of exotic substances that remain poorly understood. It is a far cry from the cosmology of the 1960s, a "pretty dismal" science, as Peebles called it, with just three numbers. Modern cosmology has been one of the great successes of Einstein's general theory of relativity and modern science as a whole, raising as many questions about the universe as it answers.

Chapter 12

═══════════

The End of Spacetime

S TEPHEN HAWKING was offered the Lucasian Professorship of Mathematical Physics at Cambridge in 1979. One of the most prestigious chairs in theoretical physics in the world, it had been held by Isaac Newton and Paul Dirac and was now being offered to a relativist not yet in his forties. Hawking deserved it. In just under two decades of research, he had made lasting contributions touching on the birth of the universe and black hole physics. His crowning achievement had been, without a doubt, the proof that black holes would radiate, had entropy and a temperature, and would ultimately evaporate. Hawking radiation had taken the world of physics by surprise. Black holes were supposed to be black and simple. Building on Jacob Bekenstein's conjecture, Hawking had shown that black holes must contain a vast amount of disorder, and that disorder is directly related to the black hole's area and not, as it is in all other familiar physical systems, its volume. The question on everyone's mind was, How is the entropy housed in a black hole? And deep down, everyone thought that quantum gravity, surely, should have the answer.

The quest for quantum gravity seemed to have come to a standstill. By the time of the Oxford symposium in 1975, when Hawking had announced his discovery of black hole radiation, it was already becoming obvious that general relativity wasn't renormalizable and that it was plagued with infinities that couldn't be hidden away. General relativity was so radically different from other theories of fundamental forces and resistant to the conventional methods that had been used to build the standard model of particles and forces. Something radically different had to be done, and Hawking and his fellow physicists faced a bewildering array of options. By the end of the 1970s, a barrage of new ideas and techniques flooded the field of quantum gravity that would cause deep rifts over the following decades. Opposing camps would cling passionately to their own set of rules on how to quantize general relativity, dogmatically refusing to accept other approaches. The community of physicists working on quantum gravity would break into opposing tribes, caught up in what some would call a veritable war. Yet, out of this turbulent and sometimes fractious environment, a common view would emerge that the old idea of spacetime as a continuum would have to be abandoned and a radically new view of reality would have to be adopted.

Stephen Hawking has always been one to make bold and controversial statements, often visionary but sometimes mischievous. On taking up the post of Lucasian Professor, Hawking used his inaugural lecture, "Is the End in Sight for Theoretical Physics?" to present his view of the future of physics, announcing that "the goal of theoretical physics might be achieved in the not too distant future, say, by the end of the century." In Hawking's mind, the unification of the laws of physics and a quantum theory of gravity was just around the corner.

He had good reasons for making his bold claim, founded in promising developments in a new idea called supersymmetry. Supersymmetry imagines a deep symmetry in nature that inextricably links all the particles and forces in the universe. Each elementary particle is supposed to have an inverse twin: for every fermion there is a twin boson and vice versa. A theory first proposed in 1976 took supersymmetry one step further and mirrored spacetime itself, creating supergravity. When Hawking gave his lecture, supergravity seemed to be the solution everyone was hoping for: a viable candidate for the quantum theory of gravity. But supergravity proved unwieldy. It extended spacetime into additional dimensions, requiring a vastly more complicated set of equations than those Einstein originally proposed. Calculating anything took months of work, and the results were plagued with infinities and particles that just didn't fit. A small group of diehards continued plugging away at it, but at least as a theory of quantum gravity, it quickly died away. Hawking would have to look elsewhere for the end of theoretical physics.

While Hawking had been optimistic in his inaugural talk at Cambridge in 1979, he had been mulling over a strange problem that he had come across while working out that black holes would radiate. This problem hovered ominously over all attempts to quantize gravity and would blow one of the most basic beliefs in physics to smithereens. Hawking would choose a meeting at the mansion of a rich entrepreneur, Werner Erhard, to foist it onto a select group of colleagues.

Erhard had made his money and fame running self-empowerment courses throughout the United States. He had been influenced by a mishmash of pundits and religions, from Zen Buddhism to Scientology, but had a penchant for physics. Every year he organized a series of lectures on physics and invited illustrious physicists such as Hawking and Richard Feynman. When, in 1981, Hawking was invited to deliver a lecture, he decided to talk about a bizarre result he had first published in 1976 that had been bothering him ever since. The talk was actually delivered by one of Hawking's young graduate students — by that point, Hawking was unable to give talks himself — and it was called the "Black Hole Information Paradox."

The talk addressed the hallowed belief in physics that given complete information about a physical system, it should always be possible to reconstruct that system's past. Imagine a ball flying by your head. If you knew how fast it was moving and its direction of flight, it would be possible for you to reconstruct exactly where it came from and what it passed along the way. Or take a box filled with gas molecules. If you could measure the positions and velocities of every molecule of gas in the box, it would be possible to determine where every particle had been at any moment in the past. More realistic situations are often much more complicated. Take the laptop I'm using to write this chapter. I would need to know a lot of information about the world to be able to exactly reconstruct how the laptop came into being, but in principle the laws of physics tell me it's possible. At an even greater level of complication, knowing all the information about a quantum state should make it possible to reconstruct the state's past. In fact, it is hard-wired into the laws of quantum physics: information is always conserved. Information is at the heart of predictability, and physicists held fast to the fundamental rule that information is never destroyed.

Information is never destroyed, that is, until it encounters a black hole. If you were to throw a copy of this book into a black hole, the book would disappear from sight. The black hole's mass and area would increase slightly, and the black hole would radiate light. Eventually the black hole will completely evaporate and disappear, leaving behind a featureless bath of radiation. If you throw in a bag of air with the same mass as the book, the exact same thing happens: the black hole's area increases, it emits light and eventually disappears, and ultimately you're left with an identical bath of radiation. The end product will be *exactly* the same in both situations even though you started off in very different ways. In fact, we don't even have to wait for the black holes to disappear. While the black holes are radiating, they will look exactly the same and it will be impossible to reconstruct whether the starting point was this book or a bag of air. Information will have disappeared.

Hawking had identified a paradox: if black holes existed, they would radiate and evaporate, but that meant that the universe was unpredictable. The idea that there was a direct connection between cause and effect, a basic assumption of Newtonian, Einsteinian, and quantum physics, would have to be thrown away. Hawking's announcement shocked his colleagues. Many of them simply refused to accept what he was saying. If information was lost, there was no future for physics as a predictive science. The only way it could be salvaged was for a black hole to be much richer than first thought, with some new type of microphysics allowing it to store information as well as making sure that by the end of its lifetime, that information is released again to the outside world. The answer could only come from quantum gravity.

• • •

In 1967 Bryce DeWitt spelled out two opposing manifestos for quantizing general relativity. Already in his forties and having spent almost twenty years trying to tackle the impossible problem, he held in his hands a trio of manuscripts summarizing his work. They became known as the "Trilogy," and to many they would become the sacred creed for quantum gravity. DeWitt was careful to acknowledge all the work that had been done on quantum gravity before him, but his manuscripts laid the foundation for marrying quantum physics and general relativity in a completely self-contained way, in essence summarizing his own work and that of everyone who had tried before him.

The first paper of the trilogy described what he called the canonical approach. It was an approach that others, including Peter Bergmann, Paul Dirac, Charles Misner, and John Wheeler, had proposed before. As in general relativity, geometry took center stage. The canonical approach breaks down spacetime into two distinct parts: space and time. General relativity stops being a theory about spacetime as one indivisible whole and becomes a theory of how space evolves in time. DeWitt then showed that it was possible to introduce quantum physics by finding an equation that can be used to calculate the *probabilities* of a given geometry of space as it evolved in time. Just as Schrödinger had done for quantum physics of ordinary systems, DeWitt found a wave function for the geometry of space.

While DeWitt would soon reject this canonical approach himself, it was quickly embraced by John Wheeler. The two met in the Raleigh-Durham airport and DeWitt shared his equation. As DeWitt recalls, "Wheeler got tremendously excited at this and began to lecture about it on every occasion." For many years, DeWitt would call it the Wheeler equation and Wheeler would call it the DeWitt equation. Everyone else simply called it the Wheeler-DeWitt equation.

The second and third papers in DeWitt's trilogy were where his heart was. They mapped out the other path, the covariant approach. In this approach, geometry was completely forgotten and gravity was just another force, carried by its messenger particle, the graviton. It was this approach that tried to mimic the successes of QED and the standard model but had led to the devastating infinities that had halted progress so dramatically at the time of the Oxford Symposium on Quantum Gravity in 1974.

The canonical and covariant approaches embodied two very different philosophies and approached the problem of quantizing gravity in two very different spirits. The canonical approach had geometry at its heart, while the covariant approach was all about particles, fields, and unification. The two approaches would pit two very different communities against each other.

The banner for the covariant approach would ultimately be carried by a radically new approach to unification called string theory. In fact, string theory started off as a cottage industry in the late 1960s, trying to explain the behavior of a whole zoo of exotic new particles that were appearing in particle accelerator experiments. The basic idea is that these particles, tiny pointlike objects, were better described in terms of microscopic, wiggly pieces of string. Particles with different masses would be nothing more than different vibrations of minute strings that floated around through space. The trick is that only one such object, one string, could describe *all* the particles. The more a string wiggled, the more energetic it was and the heavier the particle it would describe. It was a unification of sorts, but in a completely different way from what had ever been proposed.

The idea of fundamental strings was fascinating but initially flawed. Whenever anyone tried to work out physical predictions, infinite numbers kept on popping up, and they couldn't be renormalized away as in QED or the standard model. Furthermore, this theory with strings predicted the existence of a particle that behaved exactly like the graviton, the particle that was thought to be responsible for the gravitational force. While such a particle would be useful in a quantum theory of gravity, it had no place in what string theory had set out to do: explain the exotic new particles being found in accelerators.

After an initial burst of interest, string theory fell into oblivion in the mid-1970s, dismissed by most of mainstream physics. One of its few supporters, the Nobel Prize–winning physicist Murray Gell-Mann, described himself as "a sort of patron of string theory" and "a conservationist." As he recalls, "I set up a nature reserve for endangered superstring theorists at Caltech, and from 1972 to 1984 a lot of the work in string theory was done there."

In 1984, one of Murray Gell-Mann's endangered Caltech string theorists, John Schwartz, teamed up with a young British physicist from London named Michael Green. The two proposed that string theory might actually be more useful as a theory of quantum gravity. They showed how string theory in a ten-dimensional universe could incorporate quantum gravity if it satisfied certain restrictions and obeyed certain symmetries. The following year, a collective of particle physicists and relativists composed of Edward Witten from Princeton, Philip Candelas from Austin, Texas, and Andrew Strominger and Gary Horowitz from Santa Barbara, went even further. They showed that if those six extra dimensions of the universe had a very particular type of geometry known as Calabi-Yau geometry, the equations of string theory had solutions that looked exactly like a supersymmetric version of the standard model. The real standard model had to be only a short step away.

By the late 1980s, string theory had become a juggernaut. It seemed to have something for everyone. The mathematics seemed new and exciting, much as non-Euclidean geometry must have seemed to Einstein as he wielded it to understand general relativity. Mathematicians used their newest tools — not only geometry but also number theory and topology — to see what string theory could yield.

As the twentieth century came to a close, string theory hit its stride, becoming more fascinating and coherent and at the same time more complex and puzzling. At the annual string theory conference in California in 1995, Edward Witten announced that the string theory models that had emerged over the previous decade were all connected and were in fact different aspects of one underlying, much richer theory, which he called M-theory. As he put it, "M stands for Magic, Mystery, or Membrane, according to taste." Indeed, Witten's M-theory contained not only strings but also higher-dimensional objects, called membranes or branes for short, that could float around in the higher-dimensional universe.

Despite the euphoria and the hubris, string theory couldn't avoid an almost existential problem. There seemed to be too many versions of string theory available. And even if you stuck to a single version of string theory, there were many, many possible solutions that could correspond to the real world. A rough estimate led to the possible existence of 10^{500} solutions for *each version* of string theory, a truly obscene panorama of possible universes that became known as the landscape. String theory remained unable to make unique predictions.

A number of prominent skeptics argued that string theory promised too much and delivered too little. "I think all this superstring stuff is crazy and is in the wrong direction," Richard Feynman said in an interview shortly before his death in 1987. "I don't like that they're not calculating anything. I don't like that they don't check their ideas. I don't like that for anything that disagrees with an experiment, they cook up an explanation. . . . It doesn't look right."

Feynman's view was echoed by Sheldon Glashow, who, along with Steven Weinberg and Abdus Salam, had constructed the extremely successful standard model. He wrote that "superstring physicists have not yet shown that their theory really works. They cannot demonstrate that the standard theory is a logical outcome of string theory. They cannot even be sure that their formalism includes a description of such things as protons and electrons."

Daniel Friedan, a prominent string theorist in the first string revolution of the 1980s, acknowledges string theory's shortcomings. As Friedan admits, "The long-standing crisis of string theory is its complete failure to explain or predict any large distance physics. . . . String theory cannot give any definite explanations of existing knowledge of the real world and cannot make any definite predictions. The reliability of string theory cannot be evaluated, much less established. String theory has no credibility as a candidate theory of physics." These skeptics remained in the minority and were easily drowned out. If you were to enter the field of quantum gravity in the 1980s or 1990s, you might be forgiven if you thought that the covariant approach had won and string theory was all there was.

There was one thing that really riled many of the general relativists about string theory: in string theory, as in any covariant approach to quantum gravity, the geometry of spacetime, the be-all and end-all of general relativity, seemed to disappear. It was all about describing a force, like the other three forces brought together into the standard model, and how to quantize it. To a small band of relativists, the way forward was by another route, which Wheeler had embraced and De-Witt had discarded: the canonical approach. There it should be possible to cook up a quantum theory of geometry itself. In the mid-1980s, an Indian relativist named Abhay Ashtekar found a way forward.

Ashtekar was a committed relativist working at Syracuse University. He came up with an ingenious approach to untangling Einstein's field equations, rewriting them so that most of the fiendish nonlinearities disappeared and general relativity looked much, much simpler. Ashtekar's trick unlocked Einstein's equations in an unexpected way and opened the door for three young relativists to tease out their quantum nature.

Just like Bryce DeWitt, Lee Smolin fell in love with quantum gravity the moment he arrived at Harvard for graduate school in the 1970s. His adviser, Sidney Coleman, let Smolin get his hands dirty in quantum gravity by working with Stanley Deser at Brandeis. As a student, Smolin failed miserably to quantize gravity, but he remained passionate about solving the problem. It was only when he headed to Yale as an assistant professor that he realized how Ashtekar's trick made his job much easier. At Yale, Smolin teamed up with Theodore Jacobson, an ex-student of Cécile DeWitt-Morette from the Texas relativity group. Smolin and Jacobson found that instead of talking about the quantum properties of geometry at isolated points in space as they evolved over time, it was much easier to work with the geometry of a collection of points, effectively focusing on chunks of space at any given moment. In their case, the natural building blocks for the quantum theory were loops, like ribbons, in space that could be used to build solutions to the Wheeler-DeWitt equation. Things just seemed to fall into place, and a whole new way of thinking about quantum geometry emerged. The loops could link up and intertwine themselves like chain mail or an intricate fabric. As with a piece of fabric, from a distance the weaves and links disappeared and the smooth, curved spacetime of Einstein's theory would emerge. Smolin and Jacobson's approach became known as loop quantum gravity.

Smolin was joined in his quest by an iconoclastic young Italian physicist named Carlo Rovelli who had also cut his teeth working on the impossible algebra of quantum gravity. Rovelli enjoyed being a rebel. He had set up an alternative radio station during his student days in Rome, had been pursued by the Italian authorities for his political views, and had risked imprisonment for refusing conscription. Alternative views suited him. Smolin and Rovelli took the loop picture even further and looked at how the loops could be linked, braided, and knotted together. In doing so, they wandered from their starting point, the geometry of space, toward an even more broken-up and shattered view of geometry. In the mid-1990s, they stumbled upon an old idea Roger Penrose had for describing a quantum system in terms of a simple mathematical scaffolding, what Penrose called a spin network. Just like a crazy climbing frame in a children's park, the structure would be a network of links and vertices, each of which carried with it some special quantum properties. Rovelli and Smolin showed that these networks were even better solutions to the Wheeler-DeWitt equation. Yet these selfsame networks had no resemblance to the intuitive picture of space and time that any self-regarding relativist would work with.

Rovelli and Smolin's spin networks were a completely new way of looking at quantum gravity. In their model, space didn't exist at a quantum level — it was atomized or molecularized like water. Water, which looks smooth and continuous at a macroscopic level, is actually made up of molecules, little clusters of protons, electrons, and neutrons that float in empty space, loosely bound to each other through electric force. In the same way, according to Rovelli and Smolin, while space may seem smooth, it shouldn't exist if you peer at it with an extremely powerful microscope. In Rovelli and Smolin's theory, if you were able to look at distances of a trillionth of a trillionth of a centimeter, there would be no space, just the frame or network.

Loop quantum gravity was the plucky competitor to string theory in its attempts to quantize gravity. Loop quantum gravity and its progeny offered a canonical alternative to string theory's covariant approach. The devotees of loop quantum gravity made no attempt at unifying all the forces, but in taking geometry as their starting point, they tried to preserve some of the beauty of Einstein's original idea in general relativity. Ironically, in the process, they abandoned the idea of spacetime as something fundamental.

In a lecture Bryce DeWitt gave in 2004, shortly before his death, he marveled at how far quantum gravity had come along: "In viewing string theory one is struck by how completely the tables have been turned in fifty years. Gravity was once viewed as a kind of innocuous background, certainly irrelevant to quantum field theory. Today gravity plays a central role. Its existence justifies string theory! There is a saying in English: 'You can't make a silk purse out of a sow's ear.' In the early seventies string theory was a sow's ear. Nobody took it seriously as a fundamental theory. . . . In the early eighties, the picture was turned upside down. String theory suddenly needed gravity, as well as a host of other things that may or may not be there. Seen from this point of view string theory is a silk purse."

DeWitt had never worked on string theory, but it was clear where his allegiance lay. About the canonical approach he was much less enthusiastic. Despite having created it, DeWitt hated the Wheeler-DeWitt equation. He thought it "should be confined to the dustbin of history" for, among other things, "it violates the very spirit of relativity." In fact, according to DeWitt, "the Wheeler-DeWitt equation is wrong. . . . It is wrong to use it as a definition of quantum gravity or as a basis for refined and detailed analysis." He acknowledged Abhay Ashtekar's work on the equation as "elegant," but, he said, "apart from some apparently important results on so-called 'spin foams' I tend to regard the work as misplaced." DeWitt's antipathy reflected the popular view in the world of theoretical physics: string theory was winning.

The string theorists revel in what they perceive as their success. Mike Duff, now back in London, declares, "We have made tremendous progress with string and M-theory. . . . And it is the only attempt at unification." Many string theorists are convinced supersymmetry and extra dimensions will soon be discovered and that string theory is the only acceptable approach. Stephen Hawking himself has said that "M-theory is the *only* candidate for a complete theory of the universe." When asked about the rival canonical approach, seen by many as the rightful heir of Wheeler's philosophy of quantizing geometry, Duff accuses them of claiming that "quantum gravity" is synonymous with "loop quantum gravity." Duff is not alone. "They can't even calculate what a graviton does. How are they ever going to know that they are right?" argues Philip Candelas, who is firmly entrenched in the string theory camp.

In the mid-2000s, the deep-rooted antagonism between the different camps in the quest for quantum gravity came out into the open. For years, the odd op-ed articles by a few outspoken pundits had been cropping up in blogs and popular physics magazines questioning the hegemony of string theory in theoretical physics. Around 2006, two books came out claiming that string theory was, in fact, destroying the future of physics. The authors, Lee Smolin, one of the champions of loop quantum gravity, and Peter Woit, a mathematical physicist at Columbia, claimed that impressionable young physicists were being lured into working in a field that, after almost thirty years, had yet to deliver tangible hard results that would unify the forces and explain quantum gravity. According to them, academia was dominated by string theorists who hired more string theorists and kept out bright young people who didn't toe the party line. As Smolin put it in 2005, "A lot of people are frustrated that this community that styles itself as dominant — and is dominant in many places in the U.S.— is uninterested in other good work. Look, when we have quantum gravity meetings, we try to invite a representative from each of the major opposing theories, including string theory. It's not that we're so very moral; it's just what you do. But at the annual international string theory meeting, they've never done this." The blogosphere blazed with the debate while the pro–string theory camp, flustered by the attacks, took it upon themselves to set the record straight. Statements posted on physics websites were followed by hundreds of comments, a messy mélange of technical details, punditry, and pure ignorance. Everyone had an opinion.

The hostility toward string theory was palpable in 2011 when Michael Green, who had replaced Stephen Hawking as the Lucasian Professor in Cambridge, came to give a public lecture on string theory at Oxford. Green had, with John Schwartz, kick-started string theory's growth in 1984, and I had seen him give a colloquium in London in the early 1990s to enormous acclaim. String theorists were riding high then. This time, at Oxford, the atmosphere was much cooler. While most of the questions were about the specifics of his talk, a few were needling jibes. No public string theory talk can now get by without the inevitable question: "Is this theory testable?" The question always comes from someone sympathetic to the anti-string camp.

It is too early to tell how the antagonism between the different tribes working on quantum gravity will play out. For a while those working on non-string formulations of quantum gravity found it difficult to thrive, but it now seems that string theorists working on quantum gravity are being hounded, too.

A remarkable result of the debate has been that many more people are familiar with the idea of quantum gravity than before. The war between the canonical and covariant approaches has even made network TV. On the popular show *Big Bang Theory*, two characters broke off their relationship because they couldn't agree on which approach to teach their children. As Leslie Winkle says to Leonard Hofstadter as she storms out of the room, "It's a deal breaker."

Thirty years after Stephen Hawking predicted the end of physics and then unleashed his black hole information paradox on an unsuspecting world, there isn't an agreed-upon theory of quantum gravity, let alone a complete unified theory of all the fundamental forces. Yet, despite the acrimony in the quest for quantum gravity, there is common ground. A radically new and almost *shared* view of the nature of spacetime is emerging. From string theory to loop quantum gravity to all the other niche attempts at quantizing general relativity, almost all approaches give up on spacetime as something truly fundamental. This insight can be directly related to Hawking's discovery of black hole radiation and may help resolve the problem of information loss in black holes and the end of predictability in physics. One of the key steps in resolving Hawking's paradox is to understand how black holes actually store the information that they gobble up and how they might release it to the outside world. This requires a more complicated black hole than general relativity's naive picture of a horizon and nothing else. Somewhat surprisingly, both loop quantum gravity and string theory, as well as other more esoteric and more marginalized proposals for quantum gravity, seem to shed light on this problem.

In loop quantum gravity, spacetime is atomized and there is a minimum size below which it makes no sense to even talk about the concepts of area and volume. Lee Smolin, Carlo Rovelli, and Kirill Krasnov from Nottingham University have each shown how this theory makes it possible to subdivide the area of the black hole into microscopic pieces, each of which stores a bit of information like a screen of digitized information. According to the champions of loop quantum gravity, it all adds up exactly to give the right entropy of the black hole.

The string theorists see things slightly differently. Andrew Strominger and Cumrun Vafa from Harvard have shown that with M-theory, the current incarnation of string theory, it is also possible to derive an exact relationship between the entropy, information, and the area of a black hole. For a particular type of black hole they were able to show how assembling particular types of branes together allows the black hole to store just the right amount of information. The branes gave black holes exactly the right microstructure to solve Hawking's paradox. More generally, they believe that a black hole can be seen as a seething mess of strings and branes, like a tangled ball, with the ends and edges flailing about on the horizon. These bits of branes and string that bounce around on the horizon can be used to reconstruct all the information contained in the black hole. And, again, the numbers add up to give the right entropy.

While radically different, both loop quantum gravity and string theory seem to be on the right track to solve the information paradox. For, if the information actually lives on the horizon, it can feed the Hawking radiation that the black hole gradually emits, releasing information to the outside world as the black hole slowly glows. And so, by the time the black hole finally evaporates, it will have released all the information that it originally sucked in and no information will have been lost.

The string theorists are even bolder and more adventurous and claim that what they have found about Hawking radiation is an even more profound property of physical theories. Black holes seem odd because the amount of information that a black hole can store, while related to its entropy, is actually a function of its area, not its volume as one might naively expect — indeed, Bekenstein and Hawking had already argued that was so in the mid-1970s. But this means that, more generally, the maximum amount of information that can be stored in *any* volume of space will always be bounded. To find what that maximum amount of information is, just take a hypothetical black hole that contains *exactly* that volume of space and work out how much information can be stored on its surface. And so, instead of having to describe physics in a chunk of space, it should be enough to determine what happens on a surface that encompasses it, much as a two-dimensional hologram can encode all the information of a three-dimensional scene. But if this is true for a piece of space, it should be true everywhere, for the whole of the universe. In such a holographic universe, the details of what spacetime is doing at each point in the universe become irrelevant. This property is so striking that it has led Edward Witten and some of his string theory colleagues to argue that spacetime is an "approximate, emergent, classical concept" that doesn't have meaning at the quantum level. It seems that for any of the approaches to quantum gravity, at the most fundamental level spacetime might not actually exist.

When, in the 1950s, John Wheeler and his students started thinking about spacetime and the quantum, he speculated that if one were able to look really closely at space, with an inconceivably ultra-powerful microscope, one might see that "geometry in the small would seem to have to be considered as having a foam-like character." He was remarkably prescient, but from what we are beginning to understand, even Wheeler of all people might have been too conservative. Not even a foam begins to capture the complexity of where spacetime comes from.

It looks as if one of the main ideas that underpins Einstein's great theory, the geometry of spacetime itself, needs to be revisited. The quantum seems to push general relativity beyond what it is capable of describing, and a completely new way of thinking may need to be

Chapter 13

A Spectacular Extrapolation

I HAD JUST GIVEN my lecture and now stood with the audience in the atrium of the Institute of Astronomy at the University of Cambridge drinking cheap wine out of plastic cups. We gathered in small clusters, shuffling our feet, trying to fan conversation into life. The talk I had been invited to deliver that day had been about modifying gravity, describing a class of theories that proposed to dethrone general relativity as an explanation for some cosmological conundrums. The lecture itself had been uneventful. Early on, I had stumbled in refuting a comment about dark matter but had thankfully recovered. No one had told me I was wrong, nor had the questions dragged, and I was now ready to head home to Oxford.

The institute's director, George Efstathiou, strode up, eyes gleaming, brandishing his white plastic cup like a weapon. "Thank you for coming," he said. "That was an interesting talk. In fact I would say it was a good lecture about a really crap subject." I smiled politely as he slapped me on the back. It wasn't the first time I had faced this reaction and I wasn't surprised. Efstathiou had been instrumental in working out the details of how dark matter might have evolved in the formation of large-scale structure. He had also been one of the first to claim that there was evidence for a cosmological constant in the distribution of galaxies. Having risen fast in his career, Efstathiou was successful and confident. "When I took over the institute, I tried to declare it a zone free of modified gravity. And on the whole, I think I have been pretty successful." He beamed as the small group of people around us looked down at the ground. "Why on earth do you work on it?" he asked me, not really expecting an answer.

A few months earlier, I had attended a small workshop at the Royal Observatory in Edinburgh entirely devoted to discussing alternative theories of gravity. The crowd that day had included a strange mix of astronomers, mathematicians, and physicists. This meeting was different. Whenever a speaker finished a presentation, there was a round of warm applause of a kind common to self-help groups. There was also a buzz in the air, as if all the talks that day were groundbreaking revelations of some divine law of physics. Everyone was a prophet. Everyone was Einstein. The camaraderie reminded me of my brief flirtation with a Trotskyite organization in my youth, when I had experienced a heady sense of community as my fellow agitators and I agreed implicitly with one another on the innate corruption of the world.

The evangelical zeal of the workshop made me deeply uncomfortable, part of a deluded cult. After my own talk, I felt almost sickened by the applause and had to leave the room. I was being unfair; the people in that room had been working on alternative theories of gravity for years, fighting against a mainstream that believed piously in Einstein. These were scientists who would regularly have their papers rejected simply because they were about a deeply unfashionable topic. They were used to facing hostile audiences. At this meeting, their zeal fell on sympathetic ears, and they could freely discuss their goal: to overthrow Einstein's general relativity.

Most of my colleagues are reluctant to change Einstein's grand oeuvre — if it ain't broke, as the saying goes, don't fix it. Especially if you took part in the glorious renaissance of the 1960s, when general relativity had emerged from its murky, stagnant past and stepped into the limelight to become the strange, beautiful theory that could explain everything, from the death of stars to the fate of the universe. That generation of astrophysicists still feels the magical power of Einstein's theory. This depth of loyalty was made clear to me at yet another meeting, this one at the Royal Astronomical Society in 2010. In the same rooms where Eddington had announced the results of the eclipse expedition and had stamped on Chandrasekhar for invoking the specter of gravitational collapse, a gathering of astrophysicists and astronomers were asked who believed Einstein's theory was correct. A few hands went up, and a closer look revealed that these were the pioneering bunch who dragged general relativity into the mainstream in the 1960s. In the opinion of this group, general relativity was too strange and too beautiful to need changing.

No one can deny general relativity's colossal successes throughout the twentieth century, but it is due for a fresh look. Science may benefit from accepting that general relativity is going the way of Newton's theory of gravity. Newton's theory is still alive and well; it remains useful for explaining the mechanics of ballistics on Earth, the motions of the planets, and even the evolution of galaxies. The theory breaks down only in more extreme situations. Where gravity is stronger, Einstein's general theory of relativity has proved more applicable and precise. It may be time to take a further step and look for the theory that surpasses general relativity at its own extremes.

The challenges of applying general relativity on very big or very small scales, or in situations with very strong or even very weak gravity, may be indicators that the theory breaks down in some circumstances. The problematic marriage of general relativity and quantum physics may be a sign that these two theories actually behave slightly differently on the very small scales where they need to agree. General relativity's prediction that 96 percent of the universe is dark and exotic could just mean that our theory of gravity is breaking down. Now, almost a hundred years after Einstein first came up with his theory, may be a good time to reassess its true applicability.

History is full of attempts to modify general relativity. From almost the moment he published his theory, Einstein felt that general relativity was unfinished business, part of something bigger. Again and again, he tried and failed to embed general relativity in his grand unified theories. Arthur Eddington also spent the last decades of his life trying to come up with his own fundamental theory, a magical confluence of mathematics, numbers, and coincidences that could explain everything, from electromagnetism to spacetime. Eddington's quest for a fundamental theory was an endeavor that had slowly but surely eroded his prestige.

The Cambridge physicist Paul Dirac thought Einstein's general theory of relativity was the perfect example of how a theory should be. As he said in later life, "The beauty of equations provided by nature . . . gives one a strong emotional reaction," and Einstein's field equations had that beauty. Yet there was something that nagged Dirac, coincidences between numbers in nature that, if indeed the fundamental equations were beautiful, couldn't *really* be coincidences. There were some very, very large numbers in nature that couldn't be there by chance. Compare the electrical force between an electron and a proton with the gravitational force between them. The electrical force is larger than the gravitational force by a factor of one followed by thirty-nine zeros, an inordinately large number, more characteristic of a much bigger quantity, like the age of the universe. Hermann Weyl and Arthur Eddington had also argued that there must be some deep reason for the similarity of these disparate large numbers. Paul Dirac went a step further and conjectured that the strength of gravity, which is determined by a constant of nature, Newton's constant of gravitational attraction, had to evolve in time, counter to the predictions of general relativity.

Dirac proposed his idea in the late 1930s but never really took it forward. During the 1950s and 1960s Robert Dicke, one of his students, Carl Brans, in Princeton, and Pascual Jordan in Hamburg breathed new life into Dirac's idea and created an alternative to Einstein's theory. It was, to some extent, a perfect counterfoil to general relativity. As Carl Brans puts it, "Experimentalists, especially those at NASA, were effusively happy to have an excuse to challenge Einstein's theory, long thought to be beyond further experimentation." Not everyone saw it that way, and, as Brans recalls, "as time went by, many other theorists seemed also to be offended to have Einstein's theory contaminated by an additional field."

When Paul Dirac retired, he moved to Florida State University, where he indulged in some of his stranger ideas. He sometimes confided to his colleagues that he was convinced some better, more true-to-nature way of explaining gravity must exist. But he also remained wary of talking too much about his work tampering with gravity, for he felt that it would be seen by some as flaky and speculative.

By that time there had been quite a few attempts to modify general relativity, mostly driven by the problems with coming up with a good, finite theory of quantum gravity. When quantum physics is brought into the game, strange things might happen to gravity, as the Soviet physicist Andrei Sakharov pointed out in the late 1960s.

Sakharov had been part of the team, with Yakov Zel'dovich and Lev Landau and many others, that Igor Kurchatov and Lavrentiy Beria had put together to catch up with the Americans in the nuclear race. The son of a physics teacher, Sakharov entered Moscow State University in 1938 at the age of seventeen, worked through the war as a technical assistant, and finally obtained his PhD in theoretical physics in 1947. Like Zel'dovich, Sakharov emerged as a golden boy of the Soviet system. While Landau had bailed out the moment Stalin died, Sakharov had spent almost twenty years, longer than Zel'dovich, working on Soviet nuclear and thermonuclear weapons.

While Zel'dovich was creative, expansive, and intuitive, Sakharov was both more technically adept and more interested in abstract problems. The pair spoke admiringly of each other. Sakharov considered Zel'dovich "a man of universal interests," while Zel'dovich complimented his colleague's unique and idiosyncratic way of solving problems by saying, "I don't understand how Sakharov thinks."

From 1965, Andrei Sakharov focused on cosmology and gravity, but he worked at his own pace. Zel'dovich produced a torrent of papers laden with new ideas, but Sakharov was more spartan in his output. His collected works make up a slim volume. Among his meager output are some veritable gems on the formation of structure, the origin of matter, and the nature of spacetime. In one short, crisp paper, Sakharov argues that the laws governing spacetime are but an illusion and arise from the complicated quantum nature of reality. He argues that looking at spacetime and how it behaves is very much like looking at water, crystals, or other complex systems. What you think you see is really nothing more than a broad-brush picture of some more fundamental reality. The quantum properties of water molecules and how they loosely bind together are what make water look like water, a clear fluid that sloshes around and behaves the way it does. While the details differ, Sakharov's broad view proved prescient of how spacetime is perceived now, more than forty years later, as a result of progress in quantum gravity.

Sakharov looked at Einstein's theory and conjectured that the geometry of spacetime wasn't really fundamental, in the same way that the viscosity of water or the elasticity of a crystal wasn't fundamental. These were properties that emerged from a more basic description of reality. Similarly, gravity emerges from the quantum nature of matter. The surprising result in Sakharov's simple, three-page paper is that Einstein's field equations emerged naturally from such an assumption. In other words, the quantum world would naturally *induce* the geometry of spacetime. Sakharov's induced theory of gravity looked somewhat like general relativity but in fact led to a more complicated set of equations. Einstein's field equations were already a torment; Sakharov's induced gravity was far worse. The differences from Einstein's theory would really be visible only when spacetime became very curved, near black holes, or in the very early universe when everything was hot and dense, or on microscopic scales where Wheeler's quantum foam could come into play. When physical laws were pushed to extremes, they broke down and new laws emerged that encompassed the old ones.

Andrei Sakharov published his paper in 1967 when he had other things on his mind. His years working on the bomb project had brought him accolades from the Soviet regime. Like Zel'dovich, he was awarded the Hero of Socialist Labour medal three times for his pivotal role. But living up close to the bomb had made him acutely aware of the catastrophic consequences of the nuclear arms race that the Soviets were engaged in with the United States. As Sakharov increasingly objected to nuclear weapons, he also found himself losing his stature and being ignored by the regime. In 1968, he broke ranks, publishing an essay titled "Reflections on Progress, Peaceful Coexistence, and Intellectual Freedom," wherein he unequivocally declared his objections to one of the Soviet Union's main defense programs, the development of antiballistic missile defense. It was the end of Andrei Sakharov's tenure as the model Soviet citizen. The high-profile dissident was stripped of his privileges and awards, banned from working on classified projects, and exiled to Gorky. Zel'dovich frowned at what Sakharov called his "social work," saying to his closest colleagues, "People like Hawking are devoted to science. Nothing can distract them." Yet, as Sakharov wrote in his memoir, due the strength of his feelings toward the situation in the Soviet Union, "I felt compelled to speak out, to act, to put everything else aside, to some extent even science."

Sakharov may have suffered a personal setback in his scientific career, but his little idea of how the quantum might change general relativity would resurface again and again over the following decades. His paper anticipated a barrage of new quantum ideas that would batter general relativity throughout the 1970s. Some relativists thought correcting the theory in the way Sakharov had suggested would bring it more in line with the quantum world and cure the problems with infinities that plagued the theory. But by the end of the decade, Steven Weinberg and Edward Witten had proved that the infinities in such a theory couldn't cancel. Tweaking the theory wasn't enough to fix it — something more substantial had to be done.

The "super" theories — supergravity and superstrings — were definitely more substantial and seemed promising in their revisions to Einstein's theory. The fundamental idea behind general relativity remained the same — the geometry of spacetime was still center stage in the understanding of gravity. It just wasn't the four-dimensional spacetime that Einstein had originally envisaged. In the ten- or eleven-dimensional spacetimes of the super theories, the equations looked similar, but in practice the extra dimensions gave rise to a new realm of extra fundamental particles and force fields affecting the four-dimensional world that we see around us.

A few lone voices resisted this assault on general relativity, but the overwhelming feeling was that general relativity, when confronted with the quantum and in regions of high density or curvature near singularities or the Big Bang, needed to be fixed.

Einstein's theory remained a resounding success if you steered clear of the minefield of quantum gravity and didn't need to work with the universe right at its beginning, when it was hot, dense, and messy. On large scales, in astrophysics and cosmology, general relativity kept on giving.

If astronomy were an industry, the annual International Astronomical Union meeting would be its annual convention, with just about everyone trying to sell something. At the 2000 meeting in Manchester, UK, over a thousand people gathered to gloat over their recent discoveries and unveil the new projects that were about to be switched on. The cosmologists at that year's meeting were a triumphant bunch, myself included. The supernova result showing an accelerating universe had been announced a few years before. The measurements of the geometry of the universe had been announced that year. Observations were pointing to a simple yet exotic universe with dark matter and the cosmological constant. There was no more reason for disagreement and debate — personal preferences didn't matter anymore. It was good, solid science, the data was clear and consistent, and there didn't seem to be any way around it.

Jim Peebles was giving one of the plenary talks. This meeting was in some sense a celebration of Peebles's ideas and how far they had taken us. All the discoveries of the previous few years stemmed, in one way or another, from a field that he had founded with a few others. But Peebles was a staunch avoider of bandwagons, even if he was the one who set them off. In his talk, he reined in the hysteria by asking why we want to make precise measurements of the universe. And he gave his answer: to test our assumptions. He probed every angle of the Big Bang model: Why was it hot in the beginning? Where did the large-scale structure come from? How did galaxies form? In the middle of his talk, Peebles pointed out something obvious. As he later wrote in the proceedings, "The elegant logic of general relativity theory, and its precision tests, recommend GR as the first choice for a working model for cosmology." But maybe cosmologists shouldn't jump to conclusions, he warned. While general relativity had been shown to work with utmost precision on the scale of the solar system — the precession of Mercury was a beautiful example — we had no idea if we could apply it with the same level of precision on the scale of the universe. It was, he said, "a spectacular extrapolation." Peebles was right, although the conference attendees, on the whole, failed to absorb the significance of his assertion.

The French astronomer Le Verrier had argued passionately that to properly explain the drift in Mercury's orbit, there had to be a new, undiscovered planet, Vulcan, hovering at the center of the solar system. His faith in Newtonian gravity had led him to predict the existence of something new, exotic, and unseen. Without Vulcan, the Newtonian model wouldn't work. Of course, Le Verrier had been proved wrong. It wasn't a new planet but a new theory of gravity that was needed to fix the model.

Now, in the early twenty-first century, we seem to be in a similar situation, with a wonderful theory of gravity that, to explain cosmology, requires that more than 96 percent of the universe be made up of something we can't see or detect. Could this be yet another crack in the edifice that Einstein had constructed almost one hundred years before? That general relativity might have to be corrected due to quantum physics had been accepted without too much fuss. But questioning general relativity's efficacy on large scales was something different. If the dark matter and dark energy of the universe were eliminated from the picture, Einstein's beautiful theory would have to be modified. The prospect was as unappealing to many astrophysicists as taking a sledgehammer to a classic car just so it would fit in the garage.

The Israeli relativist Jacob Bekenstein started thinking about modifications to Einstein's theory in the early 1970s, while he was still a graduate student of John Wheeler at Princeton. At the same time as Bekenstein was thinking about entropy and black holes, he was also puzzled by general relativity and intrigued by the alternative theory that Dirac had proposed. "At some point," he said, "I felt I did not understand why one did things in general relativity in a certain way, why some issues were important, indeed why one followed the general path to general relativity. I felt the need to compare with a different attempt."

The "different attempt" Bekenstein chose to work on was proposed by his compatriot, the Israeli astrophysicist Mordehai Milgrom, in the 1980s. Milgrom's idea was to take a radically new look at how gravity behaved in galaxies. He pointed out that the evidence for dark matter in the rotation of galaxies seemed to arise out in the edges, where the gravitational force was very weak. If Newtonian gravity was applied in that regime of extremely weak forces, indeed it would make sense to invoke the existence of some unseen matter that could bolster the gravitational pull. But could it not be that the mistake was to apply Newtonian gravity there? So Milgrom made a bold claim, that stars well out in the tails of galaxies felt *heavier* so that the gravitational pull by the stars at the center of the galaxy on these outer stars would be much more effective than originally assumed. Because the gravitational pull was more effective, this meant the outer stars could move more quickly. This effect could explain what Vera Rubin and others had observed, that the outer parts of galaxies spin around their centers far more quickly than expected. Milgrom called his new approach Modified Newtonian Dynamics, or MOND for short.

Many astrophysicists thought Milgrom's proposal went too far in its modification of gravity. It lacked a guiding principle and passed beyond valid speculation into the realm of make-believe. When Bekenstein described the idea at an International Astronomical Union conference in 1982, he said, "Some looked at me as if I told them I have seen a UFO. . . . Almost everybody thought the emerging dark matter notion was important, and almost everybody was very much for dark matter." For the next two decades, the overwhelming majority of astrophysicists and relativists ignored Milgrom's idea or tried to shoot it down. Every now and then, a paper would apply Milgrom's law in a different astrophysical situation and show that it didn't work. Often these papers were cobbled together and incomplete, yet as long as the paper ruled out MOND, it was deemed good science and easily published. If it defended MOND, it was deemed bad science and getting it into print became an uphill struggle. MOND was, as one astronomer said, "a dirty word."

Peebles stayed above the fray, but in 2002 he did speak out on behalf of Milgrom and his ilk, chiding, "By no means have we ruled out MOND and people who do MOND should be encouraged a little more than they are." Jacob Bekenstein was more damning in his critique of how those working on MOND have been treated: "One has to take into account that the MOND versus dark matter issue is not just academic. There is a lot of money invested in searching for dark matter. . . . And no way to avoid it; entire careers are vested in dark matter. Obviously if something like MOND gets to be respectable, budgets for dark matter research will suffer and jobs will be fewer."

Since MOND's inception, Bekenstein had been trying to figure out how to make it better. Given his tendency to look at the deep roots of physical theory, he wasn't happy leaving MOND as it was. He wanted something that could be compared to general relativity and applied on all scales from the Earth all the way up to the universe. "I decided," Bekenstein said, "that it was time to meet that argument head-on by producing an example of a relativistic theory." In 2004, Bekenstein published a paper in which he constructed a new theory to rival Einstein's. He called it TeVeS, for the tensor-vector-scalar theory of gravity. It wasn't pretty. The name alluded to a jumble of fields that, when combined, led to a completely new set of field equations far more intricate and tangled than Einstein's general theory of relativity. It was a mess, but Bekenstein's theory worked. Not only did it behave like MOND when applied to galaxies, but it could be used to work out how the universe evolved and how large-scale structures formed.

The majority of cosmologists and relativists looked at TeVeS with disdain. They dismissed it as a kludge — a clumsy work-around that didn't get at the heart of the problem. Yet it was a high-powered kludge invented by a relativist with impeccable credentials. Bekenstein's black hole entropy was one of the most profound insights of modern general relativity *and* quantum physics. Yes, there was a tendency for older famous physicists to work on strange ideas, carried away by their own success. But Bekenstein wasn't like that.

Bekenstein was also not alone in his assault. While his proposal tackled the problem of dark matter, others tried to do away with the cosmological constant and dark energy. The panorama of rival theories to general relativity became messier but also richer, and the battle over the correct theory of gravity intensified. The stunning observations being made with the new telescopes and instruments developed during the explosion of physical cosmology provided additional ammunition. A pattern emerged whenever the analysis of a new piece of cosmological data was presented as confirming general relativity. The new result was inevitably tied to a press release and ensuing press coverage and, also inevitably, followed by a flurry of papers pointing out that what seemed like incontrovertible evidence for general relativity wasn't really that solid.

In January 2008 a paper in *Nature* signaled yet another quiet shift. In it, an Italian team of observers analyzed the data from a survey of galaxies. It was the kind of thing that Jim Peebles and his followers had been doing for almost forty years. By studying how the galaxies were clustered together, the Italian team was able to measure the rate at which they were falling into each other, attracted by the gravitational field in which they were immersed. This was nothing new. It had been done quite a few times before with different surveys of galaxies. But what was interesting was how they presented their results: on the graph where they presented their data, the Italians superimposed what would be expected from general relativity but also from a few other, alternative models of gravity. Some of the theoretical predictions went straight through the data points, and others missed them completely. It was an obvious thing to do: compare theory and observation.

The *Nature* paper heralded a change in spirit and emphasis among the observers in cosmology. The emphasis since the late 1990s had been solely to measure, characterize, and nail dark energy, but this paper instead used cosmological observations to test general relativity. It was a return to testing the fundamental assumptions of physical cosmology.

In the ensuing years, testing general relativity has been at the heart of observational cosmology. We still want to know if there is dark energy, what it is made of, and how galaxies assemble themselves to become the building blocks of the universe. But again and again, in requests by scientists for funding, in seminars and plenary lectures, testing general relativity has taken center stage.

Modifying gravity is still frowned upon by many if not all relativists. While tampering with general relativity when it comes up against the quantum is quietly accepted, fixing spacetime to agree with observations is something else. There is still so much to understand and discover in Einstein's theory, and for relativists, changing it is an unnecessary and inelegant complication. But nature may not agree, and with astronomers taking an interest in Einstein again, we now have an opportunity to explore the fundamental laws of spacetime, looking farther and deeper in the cosmos.

The ideas of Dirac, Sakharov, and Bekenstein, bolstered by new work in observational cosmology, offer a new way of thinking that is too exciting to ignore and give new purpose to the juggernaut of cosmology. Some of my colleagues at Oxford and Nottingham and I recently decided to write a survey of the field of modified gravity. We felt like jungle explorers uncovering new exotic species. There were dozens of theories, each one odder than the next, proposing quirky modifications to general relativity, often with surprising and realistic results. Our review presented a rich bestiary of gravitational theories, many of which could give general relativity stiff competition. There are so many people thinking about alternatives to general relativity that today's big general relativity meetings — the successors to the DeWitts' Chapel Hill conference and Alfred Schild's Texas Symposiums — offer parallel sessions packed with speakers from all generations and continents trying to take general relativity apart. It is still a fringe activity, but it's one with many activists.

When I gave my talk that afternoon in Cambridge, Efstathiou had been dismissive. Yet even Efstathiou, a brilliant mind and one of the pioneers of the current standard cosmological model in which general

Chapter 14

Something Is Going to Happen

I RECENTLY SPENT SOME time advising the European Space Agency. ESA is responsible for sending scientific satellites into space, often cooperating with NASA. One of its most famous experiments is the Hubble Space Telescope, which has been used to take some of the crispest, cleanest images of deep space.

Satellites are the new outposts of science, unspeakably sophisticated laboratories where virtually unimaginable experiments can take place, floating in space at the boundaries of our reach. And they are expensive, costing anywhere from half a billion to many billions of dollars each. You don't just chuck these beasts up into the sky. It takes years — sometimes even decades — of planning and design before a firm decision is made on whether sending them up is worthwhile.

At ESA, we discussed what humanity's future space missions should be along with various proposals that were being made by large international teams of scientists. During the long, drawn-out meetings in which we were clobbered with PowerPoint presentations, Gantt charts, and costs that made my eyes water, I would often lose the will to live. This science seemed so different from the freewheeling exploration, unbridled creativity, and beautiful mathematics that had pulled me in as a graduate student. It was also shocking that we were discussing such far-reaching, breathtaking missions as if they were corporate enterprises, like opening new factories in some faraway land.

What struck me forcibly was how, in the middle of the drudgery and technospeak, general relativity was at the heart of the scientific case for so many of the satellite missions proposed. Yes, general relativity was writ large on all the proposals, hovering magnificently above the specifics and technicalities that we were discussing. There and then, we were being asked to fund billion-dollar missions that would either test Einstein's theory or use it to explore the outer recesses of space and the inner workings of dense, massive objects. It was the future of space science in the twenty-first century. Not all the proposals could be funded, not all the satellites would fly, and the choice was breathtaking.

One mission proposed to pick out the ripples of space and time, the waves of gravity expelled from the explosive collisions between black holes. It would be the spawn of LIGO and GEO600, a humongous interferometer made up of not one but three satellites orbiting the sun with ultraprecision laser beams bouncing back and forth between mirrors spaced millions of kilometers apart. Called the Laser Interferometer Space Antenna, or LISA, it would mop up after the ground-based experiments that are currently coming online, picking up the faint signals that LIGO and GEO won't see.

That wasn't all. Another mission was proposed to measure the history of the expansion of space back to when the universe was a hundredth of its current age. It would take the methods of physical cosmology and push them to eleven, surveying swaths of the sky to build up catalogues with hundreds of millions of galaxies. Then, by looking at how the galaxies are assembled together into the vast cosmic web, carefully studying how the clusters and filaments of light came together around voids through gravitational collapse, it would be possible to figure out the effects of dark matter and dark energy or whether, indeed, as some now seem to believe, Einstein's theory breaks down on the largest scales.

There was yet another proposal for a satellite that would target the inner cores of black holes and look for the powerful x-ray emissions that had opened up such a phenomenal window on the universe in the late 1960s and 1970s. This time, it would be possible to go further and look at how the extremely warped spacetime near their centers would shred matter and light apart, just as Zel'dovich, Novikov, Rees, and Lynden-Bell had claimed it would. For the first time, it might just be possible to measure physical processes that happen close up to the infamous event horizon, the Schwarzschild shroud that had baffled so many for so long.

During those meetings it became clear to me that general relativity will be at the heart of physics and astronomy in the twenty-first century.

It's not going to be easy. The real world of tightened budgets, poverty, and recession make many think twice about spending billions of euros or dollars on a satellite mission. While it's not surprising that the US government decided to pull out of funding LISA, it's still devastating.

LISA was to be the final step in the discovery of gravitational waves. Not only would LISA discover these elusive ripples, it would be a colossal, perfect observatory for using them to look at black holes colliding and neutron stars circling each other. LISA would let us learn so much about all the fantastic exotica that Einstein's theory of relativity predicts. The first stage of LIGO was a huge success even though it didn't see anything. It proved that the technology, an insane mishmash of lasers, quantum, and precision engineering, actually works and is gearing up to be made even better. The next stage of LIGO, known as Advanced LIGO, may see something and prepare the way for LISA. But now, with the Americans pulling out, LISA is on the ropes. Who would be willing, in such a time of need, to fund a big beast with such an esoteric goal?

The quest for gravity waves is just too important to be given up. And so the Europeans, through ESA, will go ahead. The new interferometer will be smaller, but still spectacular. It will still cost billions, just not as many. And the distraught relativists in America have regrouped and refuse to give up. Quietly, a number of groups scattered throughout the country have set to work trying to come up with their own proposal for something cheaper, more compact, and less ambitious that would still be able to look into the far recesses of spacetime.

If the Europeans have a change of heart or are further consumed by the financial crisis, there will be a backup plan.

We don't have to wait for the satellites to go up. Fantastic things are already happening. We've seen the checkered history of the singularity and how repugnant it was to so many great minds, from Albert Einstein and Arthur Eddington to John Wheeler (until he saw the light). With the discovery of quasars, neutrons stars, and x-rays and the phenomenal burst of creativity from the likes of Wheeler, Kip Thorne, Yakov Zel'dovich, Igor Novikov, Martin Rees, Donald Lynden-Bell, and Roger Penrose, black holes became firmly cemented in our consciousness. By the end of the period in the 1960s and 1970s that Kip Thorne called the Golden Age of General Relativity, black holes had become real things, as much a part of astrophysics and physics as stars and planets.

On my shelf I have two textbooks on general relativity that came out at the end of the golden age. They are very different. One of them, *Gravitation,* was written by John Wheeler and two of his brilliant ex-students, Charles Misner and Kip Thorne. It is over a thousand pages long, big with a black cover like a gothic phone book, exquisitely illustrated and packed with just about everything you might want to know about spacetime. MTW, as it is known, has all the odd stuff in it, the Wheelerisms that Wheeler kept on coming up with in his talks and conferences. The other textbook is by Steven Weinberg, one of the fathers of the standard model of particle physics. While Weinberg has established himself as one of the towering intellects of the quantum, he also dabbled in general relativity, and his book *Gravitation and Cosmology* is a careful, considered introduction to Einstein's theory. It has much of what MTW has but without the madness. And, given the exciting discoveries of the decade that preceded it, Weinberg's book doesn't have much on black holes. In fact, black holes are cautiously mentioned toward the end of a subsection in the middle of the book as something to watch out for, as if black holes come from pushing general relativity just a bit too far.

You can see why some people were still cautious. Yes, all the evidence seemed to point at dense, heavy objects both far and near. And it was difficult to explain them any other way if not as black holes. But truly, no one had actually *seen* a black hole. To look at black holes directly is a bit of a paradox. There is nothing there to see — black holes are invisible behind the Schwarzschild shroud. Just because we can't see them doesn't mean they aren't worth looking at. In fact, we have a gigantic black hole sitting right in the middle of our galaxy, the Milky Way. It weighs about a hundred million times more than the sun and has a radius of about 10 million kilometers. It is big. But it is also tens of thousands of light-years away, which means that it takes up only about a hundred-millionth of a degree in the sky, making it tinier than a pinprick from our point of view, far smaller than we are able to resolve with our current telescopes. It is only through the cleverness and perseverance of astronomers that we can be assured a black hole is there.

Two research groups, one based in Munich and another in California, have patiently followed the motion of a few stars that are hovering close to the center of the Milky Way. Over more than a decade they have been able to track the motion of that group of stars as they swoop around and around, and they have found that the stars move in incredibly curved orbits, clearly being pulled in by some gigantic gravitational force. By carefully measuring the orbits of these stars, they are able to figure out not only how strong gravity is in that region but also where all that gravitational pull is coming from. Combining these observations, the two groups are able to measure the mass of the black hole with exquisite precision and to pin down where the singularity in spacetime should be.

There is more. Astronomers and relativists are mobilizing to build the telescope that will actually *see* the black hole. Called the Event Horizon Telescope, it will have a resolution of a billionth of an angular degree, a fraction of the size the black hole takes up in the sky, so that it will actually be able to see Schwarzschild's shroud, the surface of the black hole that Oppenheimer and Snyder showed is a snapshot frozen in time. It will be a dark shadow surrounded by the swirling mess that Zel'dovich and Novikov conjectured would surround the black hole, the accretion disk of stars, gas, and dust being shredded by the gravitational pull of the singularity.

The accumulating evidence is very compelling. While Weinberg's reticence was understandable, it is now difficult to find anyone who would argue against there being a black hole in the center of the Milky Way. And just like the Milky Way, all the other galaxies should have black holes firmly in their centers like massive engines surrounded by gargantuan spirals of stars.

The media find anything to do with general relativity and Einstein's great ideas both enticing and newsworthy. Images of the center of our galaxy lead to headlines like "Black Hole Confirmed in Milky Way" on the BBC, and "Evidence Points to Black Hole at Center of the Milky Way" in the *New York Times*. On the day I am writing this, the BBC news website features a comment from an Oxford colleague of mine on a recent observation of a quasar now shown to be a super-massive black hole with a mass of a billion suns. What stuns me is that almost fifty years after Maarten Schmidt's measurements and the first Texas Symposium, black holes can still create such a stir.

Not a month goes by without something in the news about cosmology or black holes, about the beginning of the universe or echoes of other universes, signatures of the mysterious multiverse. Words like *black holes, Big Bang, dark energy, dark matter, multiverse, singularity,* and *wormholes* have penetrated the farthest recesses of popular culture, from Broadway plays and songs to comedy shows and Hollywood movies. And then there are the countless ways in which general relativity has been folded into science fiction, from novels to TV shows to movies. They have surpassed even Wheeler's wildest dreams in terms of imagination and creativity. Everyone seems to consider himself or herself an expert on general relativity.

This fascination is exhilarating but sometimes also ludicrous. When my son called me irresponsible for, in some indirect way, willing the Large Hadron Collider into existence, he was not alone. The media had repeatedly advertised the idea that string theory, one of the candidate theories of quantum gravity, predicted that black holes would be formed when the Large Hadron Collider switched on. When the beams of protons actually collided, among the multitude of stuff that would spew out into the detectors would be microscopic black holes, mini portals into other dimensions. My son also knew that black holes suck up everything around them. Everyone knows that. So why on earth would I, or anyone in his or her right mind, want to produce these incredibly dangerous things? It was obviously a stupid thing to do.

One physicist of sorts, of all people, actually tried to stop the LHC from being switched on by going to court. When interviewed on the Jon Stewart show, he was asked about the probability of a catastrophe actually happening, and in a remarkable flourish of on-air reasoning he said, "Fifty percent." He lost, the LHC was switched on, and we are still here. Unfortunately, no miniature black holes have been found.

Every time I give a public lecture about what I do, I am asked the same thing: "What was there before the Big Bang?" I resort to the various explanations. There is the "There was no before, no time, before the Big Bang" answer. Or there is my colleague Jocelyn Bell Burnell's more Zen-like answer: "That is like asking what is north of the North Pole." It would be so much easier if I could resort to mathematics, but I can't because most of my audience would find that it went over their heads. And for decades, because of Stephen Hawking's and Roger Penrose's singularity theorems, we have believed that, indeed, there was nothing before the Big Bang. It is one of those truths, those *mathematical* truths, we can't get around that came out of the Golden Age of General Relativity.

Very recently, I've found my answers to the Big Bang questions becoming much more diverse and much less definitive. Over the past few years, the beginning of time has been thrown wide open by developments in quantum gravity and cosmology. When you wind back the clock and make the universe denser, hotter, and messier, that is when quantum foam, strings, branes, or even the loops have a say. It is where, it seems to some, spacetime breaks down and it no longer makes sense to talk about the initial singularity.

So what happened before the Big Bang? One possibility is that our universe popped into existence out of a vacuum, a bubble of spacetime that grew and grew to become what we are today. And like ours there are many universes that just popped up out of the vacuum. Another guess comes out of ideas in string and M-theory, which posit that the universe has many more than four dimensions and that we live on a three-dimensional "brane" in this spacetime and roll with it. Our domicile, our brane, feels just like a three-dimensional universe that every now and then collides with another brane just like ours. When they collide, they heat up, and as a result our universe feels as if it has undergone a hot Big Bang. There is no singularity, just an infinite succession of hot Big Bangs, a cyclic universe that would have made the Soviet orthodox philosophers, and possibly even Fred Hoyle and his cronies, proud. The model's creators have dubbed each new Big Bang *Ekpyrosis*, an ancient Greek term for the periodic destruction of the universe, inevitably followed by a rebirth.

But, of course, so much of quantum gravity seems to point to the fragmentation of spacetime if looked at under an all-seeing microscope. If we wind back the clock so that spacetime is concentrated at a point, surely we must run up against the bits and pieces that make up the fabric of space. Before any initial singularity is reached, when the chunkiness comes into play, known physics breaks down. Those who believe that loop quantum gravity is the answer say that there was a before, a time when the universe was collapsing until it reached the quantum wall and magically started expanding again. The universe underwent what has prosaically become known as a "bounce."

It might not even be necessary to resort to that weird, dark era where quantum gravity comes into play, where so many differing opinions lead to so many different conjectures. A grander possibility is that spacetime is much vaster than we previously envisioned and our universe is only one of countless universes that together make up the multiverse. All over the multiverse, universes are breaking out into existence, growing to cosmic proportions, each one at its own pace and made up in its own particular way. If we follow back the existence of our own universe, we find that it is embedded like a pustule in a much wider spacetime that has existed for all eternity. The multiverse is a wild, immense realm of what is ultimately stasis: a steady state of creation and destruction.

The multiverse, along with something called the anthropic principle, has emerged as the favorite solution for the cosmological constant problem. With the great successes of observational cosmology, many believe that the cosmological constant actually exists in the real universe, even though quantum theory predicts an obscenely large value for it, much larger than what we observe. String theorists now apply the lack of predictivity in string theory to posit a landscape of different possible universes, each one with its own symmetries, energy scales, types of particles and fields, and, most crucially, its own cosmological constant. Any of these universes is possible, even ones with a very small cosmological constant. The anthropic principle, first proposed by Robert Dicke and further developed by Brandon Carter, argues that the universe is the way it is because if it were any other way, we wouldn't be around to see it. We exist and are sentient because the universe has exactly the right set of constants, particles, and energy scales — including the cosmological constant — that allow for our existence. There are countless possible universes, but only the ones with the right values for physical constants, including the cosmological constant, allow us to exist. Given that such a universe is possible, it is natural that it will be the one, of all the universes in the multiverse, that we observe.

Some argue that cosmology has become so rich and complex that we may be at the frontier of what should be called science. George Ellis is one skeptic who thinks this approach goes too far. A relativist who, with Hawking and Penrose, cemented the existence of singularities in the cosmos in the late 1960s, Ellis has been at the forefront of using the whole of the universe as an immense laboratory and testing ground for Einstein's theory. "I do not believe the existence of those other universes has been proven — or could ever be," he says. "The multiverse argument is a well-founded philosophical proposal but, as it cannot be tested, it does not belong fully in the scientific fold." On this landscape of possibilities anything can be predicted somewhere. Even among the string theorists there is a sense that things have gone too far. The new approach abandons the ultimate goal of modern physics to find a unique and simple unified explanation for all the fundamental forces, including gravity. Accepting the multiverse is tantamount to giving up. Even Edward Witten, the pope of modern string theory, is unhappy with how things are turning out and says, "I hope the current discussion of the string theory isn't on the right track."

Yet the multiverse's following is growing. It solves some of the great unsolved problems, such as why there is a cosmological constant and why the constants of nature are tuned to be exactly what we measure them to be. On a regular basis, there are press releases and media reports on parallel universes and evidence for the immensity and plurality of spacetime. It is, of course, a wonderful setting for speculation, a vast blank canvas for storytelling. But, to Ellis, it simply isn't science.

In 2009 I visited Príncipe, a small, lush speck of greenery in the armpit of Africa. It was from there that, ninety years before, Arthur Eddington had telegraphed a message to Frank Dyson, then the president of the Royal Astronomical Society, saying simply, "Through cloud. Hopeful." Eddington's measurements of starlight during a solar eclipse had established Einstein's general theory of relativity as *the* modern theory. The eclipse expedition established Eddington and Einstein as international superstars.

I traveled to the small island nation of São Tomé and Príncipe with a motley collection of Brits, Portuguese, Brazilians, and Germans to lay a plaque donated by the Royal Astronomical Society and the International Astronomical Union at the site where Eddington and Cottingham had made their measurements.

São Tomé and Príncipe had emerged from centuries of colonial rule to become, for a while, yet another African socialist state. It joined the world of free markets, and its jumbled collection of shiny new houses for affluent Angolan holidaymakers contrasted with grand, decrepit colonial farmhouses.

The main house at Roça Sundy, where Eddington made his measurements, was in better shape than most of the abandoned colonial homes scattered throughout the green countryside. The regional president of Príncipe, a tiny island of not more than five thousand people, had taken it as his holiday home. This turned out to be wishful thinking — it was still ramshackle, rusty, and uninhabitable.

I found that perfect little corner of the world deeply moving. My grandmother was born in São Tomé and Príncipe in the early twentieth century, and I had heard much about the place from her. But more important, I felt that I was witnessing a turning point in history. This is where Einstein's theory was proved right, insofar as any scientific theory can be proved right. This is where general relativity became fact.

Scattered around were relics of the bygone era when Eddington passed through. There was the tennis court, cracked concrete fighting a losing battle with the inexorable vegetation seeping up from the ground. Everywhere I looked was lush, overwhelming green. It was a far cry from the bleak, manicured landscape of the fens where Eddington had spent almost all his life. Now, with our visit, there was a shiny plaque marking Eddington's achievement and, we hoped, explaining to any passerby of this remote location how stupendous the event had been.

Looking back to 1919, it is amazing how Einstein's and Eddington's ideas developed. The simple idea that light would be deflected by warped spacetime, the key to testing Einstein's theory, was now, ninety years later, one of the most powerful tools in astronomy. Over the past twenty years it has become the norm to look at how light is deflected by spacetime to learn about the universe. By looking at stars in nearby galaxies and waiting to see if their light is suddenly focused due to the passage of a dark heavy object in front of them, it has been possible to look for dark matter in our galaxy. The nuggets of dark matter, if they exist, will play the role of the sun in Eddington's experiment, bending starlight, lensing it, as the effect has become known. On a grander scale, we now use lensing to look at clusters, swarms of tens to hundreds of galaxies. These behemoths sink into spacetime, creating gigantic warps that scatter and align the light from distant galaxies. Astronomers now use the distortions and shifts in the light of these distant galaxies to weigh the clusters.

Why stop there? With typical hubris, astronomers, cosmologists, and relativists have now set their sights on mapping the distortions of spacetime all the way out, as far as can possibly be observed. By observing slices of the universe and seeing how the light of those galaxies is affected by intervening spacetime, it should be possible to build up a detailed description of what spacetime actually looks like all around us. Taking Einstein's and Eddington's ideas to a new level, we harness the universe, learning what it is made of and whether our current laws for the way spacetime behaves are correct.

Throughout the day, as festivities continued in Príncipe, Einstein's and Eddington's names were on everyone's lips. In this lost corner of a minuscule island, it was too much to ask that anyone would actually know what we were talking about. Ponderous nods from the local and visiting dignitaries didn't mean much, and a shoal of children and teenagers ran around during the ceremony. They didn't know what it was about, true, but they had of course heard of Einstein. And some even knew about the famous Englishman Eddington who had come to visit many years ago. They all agreed that it was a good thing — that small island's claim to fame.

As I watched the crowd joining in this odd, esoteric celebration, I saw it as yet another quirky sign of how universal and democratic Einstein's theory has become. While tortuous and often intractable, Einstein's theory has been at the same time democratic, easily encapsulated in a few pages of condensed equations. The history of general relativity spans many continents, with a full cast of characters that is truly international and varied. British astronomers, a Russian meteorologist, a Belgian priest, a New Zealander mathematician, a German soldier, an Indian child prodigy, an American expert on the atom bomb, a South African Quaker, and so many more have been brought together by the elegance and power of Einstein's theory.

That night, we handed out telescopes to the crowd and looked up at the stars. The sky was breathtaking, ready to offer up much more that would help us delve deeper into Einstein's theory. I thought of how, even now, Einstein's theory was driving us to look out into the cosmos on a grander scale. The new Príncipe might now be in the south of Africa or in the Australian desert, and the new telescope would use the latest, most powerful technologies of the twenty-first century.

While Eddington had used an optical telescope, something with a lens, an eyepiece, and a photographic plate, this new phase will rely on radio antennas and dishes. Radio has already given so much to general relativity, but this time it will go much further than has ever been envisaged. The idea is to build a collection of tens of thousands of radio antennas scattered across hundreds and thousands of kilometers. Known as the Square Kilometer Array, or SKA, because the total collecting area of all the antennas should add up to a square kilometer, it will take one, possibly two continents to support it. Some of the telescopes will lie out in the vastness of the Australian west, and others will be strewn throughout southern Africa. The core of the beast will be laid out in the Karoo Desert, but a number of these dishes will be scattered throughout the continent in places like Namibia, Mozambique, Ghana, Kenya, and Madagascar. It will be a truly continental, *African* endeavor. And, in the same way that Eddington used Príncipe to establish general relativity, the SKA would be the beast that could test Einstein's theory on cosmological scales with unprecedented precision. The SKA would detect if there were, indeed, any cracks in Einstein's grand idea. It would be able to detect the elusive gravitational waves that are still out there, waiting to be discovered. It might even reveal the nature of the infamous dark energy that seems to have cemented itself into the current model of the universe.

That night as we celebrated Eddington's and Einstein's colossal achievements, I thought about how we are only at the beginning of what the theory of spacetime is going to tell us about the universe. The twenty-first century is surely going to be the century of Einstein's general theory of relativity, and I feel fortunate to be living at a time when so many new things are waiting to be discovered. Almost a hundred years after Einstein finally came up with his theory, something fantastic is going to happen.

Acknowledgments

Two people made this book happen. Patrick Walsh convinced me, and gave me the opportunity, to write about this obsession of mine. Courtney Young took my manuscript and, with remarkable grace and firmness, made it into something I would want to read.

I have relied on testimony, advice, and criticism from a long list of colleagues, friends, family, readers, and writers over many years. Here is an attempt at a (quite possibly incomplete) list: Andy Albrecht, Arlen Anderson, Tessa Baker, Max Bañados, Julian Barbour, John Barrow, Adrian Beecroft, Jacob Bekenstein, Jocelyn Bell Burnell, Orfeu Bertolami, Steve Biller, Michael Brooks, Harvey Brown, Phil Bull, Alex Butterworth, Philip Candelas, Rebecca Carter, Chris Clarkson, Tim Clifton, Frank Close, Peter Coles, Amanda Cook, Marc Davis, Xenia de la Ossa, Cécile DeWitt-Morette, Mike Duff, Jo Dunkley, Ruth Durrer, George Efstathiou, George Ellis, Graeme Farmelo, Hugo and Karin Gil Ferreira, Andrew Hodges, Chris Isham, Andrew Jaffe, David Kaiser, Janna Levin, Roy Maartens, Ed Macaulay, João Magueijo, David Marsh, John Miller, Lance Miller, José Mourão, Samaya Nissanke, Tim Palmer, John Peacock, Jim Peebles, Roger Penrose, João Pimentel, Andrew Pontzen, Frans Pretorius, Dimitrios Psaltis, Martin Rees, Bernard Schutz, Joe Silk, Constantinos Skordis, Lee Smolin, George Smoot, Andrei Starinets, Kelly Stelle, Francesco Sylos-Labini, Kip Thorne, Neil Turok, Tony Tyson, Gisa Weszkalnys, John Wheater, Adam

Wishart, Lukas Wilowski, Andrea Wulf, and Tom Zlosnik. While their contributions have been invaluable, any errors or misconceptions in the final text are my own.

The team at Conville and Walsh have been incredibly supportive in seeing this book through, and my colleagues at the University of Oxford have been enthusiastic and helpful. It is a real privilege working with them all.

Notes

One of the joys of writing this book has been reading many of the original papers and articles on general relativity as well as histories, biographies, and memoirs. I hope the specific sources that follow will be taken as encouragement for further reading in the subject. It is definitely worth the effort. Full references for the publications cited in this section can be found in the bibliography.

I highly recommend plowing through some of the scientific literature, even if you don't have the background to understand much of what is being done. It will give you a real flavor of what science is about, how things are presented, explained, and promoted, and how the vast cast of characters interact with each other through the scientific journals. Unfortunately, many of the journals are behind "paywalls," and some of the articles I refer to here cannot be accessed if you are not in an academic institution. A surprising number of them can, however, and I suggest you look for them. I recommend using one of the following search engines:

http://scholar.google.com

http://inspirehep.net

http://adsabs.harvard.edu/abstract_service.html

Each one has its own syntax, but collectively they will help you find any of the articles you might be looking for. The scientific community in astronomy, mathematics, and physics has, for the past two decades, posted freely available copies of articles at a repository on http://arxiv.org. Wherever possible, I have listed the link a given paper has to that website.

Finally, I interviewed a few of the protagonists for this book; in the notes that follow, I explicitly identify quotes that came from those interviews.

PROLOGUE

The description of A. Eddington's encounter with L. Silberstein is described firsthand in Chandrasekhar (1983). You might want to venture onto the "gr–qc" section of ArXiv.org to see the kind of weird but sometimes wonderful stuff that pops up in the field of relativity.

1. IF A PERSON FALLS FREELY

So much has been written about Einstein that I have been spoiled for choice. I have used a handful of superb biographies to guide me through his life. Fölsing (1998) is very detailed, nuanced, and richly documented. Isaacson (2008) captures the essence of the man, bringing real color to his life and times. Pais (1982) is a classic, focusing on his work and mapping out many of the mathematical and physical steps that led to his great discoveries.

As a panorama of physics at the beginning of the twentieth century, there is Bodanis (2001), a wonderful piece of narrative history, focusing on the lead-up to, and consequences of, Einstein's famous $E = mc^2$. Bodanis (2006) offers real insight into how Maxwell and his contemporaries transformed the world with their work on electricity and magnetism. Baum and Sheehan (1997) walk us through the beginning of the end of Newtonian gravity and Le Verrier's ill-fated quest for the planet Vulcan.

There is a whole world of Einstein scholars out there. John Norton, John Stachel, and Michel Janssen, to name a few, have all really tried to get into his mind, spelling out his successes and failures. It is a rich literature that can really suck you in. Those who want a firsthand look at his discoveries, especially his miraculous year of 1905, should have a look at Stachel (1998), a compilation of his papers. Einstein's first step in his quest for general relativity, the article for the *Yearbook,* is well worth a look, but it is probably easier to read a more gentle description in Einstein (2001).

Page

1 "When you pick up an application": F. Haller, in Isaacson (2008), p. 67.

2 "You are a very clever boy": H. Weber to Einstein, in Isaacson (2008), p. 34.

3 "considerably facilitates relations": Einstein to W. Dällenbach, 1918, in Fölsing (1998), p. 221.

5 "asymmetries": Einstein in Stachel (1998) and Pais (1982), p. 140.

7 Proust and Le Verrier: See Proust (1996).
   Dickens and Le Verrier: See Dickens (2011).

8 "How could a planet": Le Verrier, 1859, in Baum and Sheehan (1997), p. 139.

9 "If a person falls freely": Einstein lecture in Kyoto, 1922, in Einstein (1982).

11 "My papers are meeting with much acknowledgement": Einstein to M. Solovine, 1906, in Fölsing (1998), p. 201.

"I must confess to you that I was amazed": J. Laub to Einstein, 1908, in Fölsing (1998), p. 235.

2. THE MOST VALUABLE DISCOVERY

While Fölsing (1998) does a careful job of describing the context for the discovery of general relativity and how Einstein stumbled toward his final version, Pais (1982) provides the detail — the latter is very mathematical but also very rewarding. For Eddington I have relied heavily on three very different books. Chandrasekhar (1983) is a slim, respectful volume on his work and thought. Stanley (2007) addresses his more mystical and political stance and how he behaved during the First World War. Miller (2007) is a fantastic read where we get a sense of how complex Eddington was (and how difficult he would become later in life). A careful description of the eclipse expedition can be found in Coles (2001).

12 "You know, once you start calculating": Fölsing (1998), p. 311.

"mathematically cumbersome": H. Minkowski to his students, in Reid (1970), p. 112, and Fölsing (1998), p. 311.

"superfluous erudition": Fölsing (1998), p. 311.

"Since the mathematicians pounced": Ibid., p. 245.

14 "You've got to help me": Ibid., p. 314.

15 "The gravitation affair has been clarified to my full satisfaction": Einstein to P. Ehrenfest, in Pais (1982), p. 223.

16 "in the madhouse": Einstein to H. Zangger, 1915, in Fölsing (1998), p. 349.

"the life or property": Fölsing (1998), p. 345.

17 "educated men of all states": Ibid., p. 346.

18 Meeting C. Perrine: Mota, Crawford, and Simões (2008).

"We can readmit Germany to international society": H. Turner, 1916, in Stanley (2007), p. 88.

19 "Think, not of a symbolic German": Eddington (1916).

22 "there has been between us something like a bad feeling": Einstein to D. Hilbert, 1915, in Fölsing (1998), p. 376.

"the most valuable discovery of my life": Einstein to A. Sommerfeld, 1915, in Fölsing (1998), p. 374.

23 "We have tried to think that exaggerated and false claims made by Germans": H. Turner, 1918, in Stanley (2007), p. 97.

"under present conditions the eclipse will be observed by very few people": F. Dyson, 1918, in Stanley (2007), p. 149.

24 "Through cloud. Hopeful": Pais (1982), p. 304.

25 "Eclipse Splendid": Ibid.

"the most important": J. J. Thomson, 1919, in Chandrasekhar (1983), p. 29.

26 "Revolution in Science": *The Times,* November 7, 1919.

"All Lights Askew": *New York Times,* November 10, 1919.

27 "In Germany I am called a German man of science": Einstein on his theory, *The Times,* November 28, 1919.

3. CORRECT MATHEMATICS, ABOMINABLE PHYSICS

There is a wealth of information about the discovery of the expanding universe. The main papers can be found in the compilations of cosmological classics, a notable example of which is Bernstein and Feinberg (1986). I have avoided all discussion of "Mach's principle," which pushed Einstein to formulate his static universe model, but you can find a discussion of the debate between Einstein and de Sitter in Janssen (2006). A detailed and well-documented history of the expanding universe is Kragh (1996) and more recently Nussbaumer and Bieri (2009). For individual and more detailed descriptions of the main protagonists in this chapter see Tropp, Frenkel, and Chernin (1993) for Friedmann, and Lambert (1999) and the article by A. Deprit in Berger (1984) for Lemaître. An entertaining description of Hubble and Humason can be found in Gribbin and Gribbin (2004), and the Humason AIP interview in Shapiro (1965) is hugely informative. For some of the controversy over who did what in the discovery of the expanding universe (and the underappreciated role that Vesto Slipher played) I recommend Nussbaumer and Bieri (2011) and Prof. John Peacock's homage to Slipher at http://www.roe.ac.uk/~jap/slipher.

30 "The introduction of such a constant implies a considerable renunciation": Einstein (2001).

"committed something in the theory of gravitation that threatens to get me interned in a lunatic asylum": Einstein to P. Ehrenfest, 1917, in Isaacson (2008), p. 252.

31 "To admit such possibilities seems senseless": Ibid.

33 "The cosmological constant . . . is undetermined": Friedmann (1922), reprinted in Bernstein and Feinberg (1986).

34 "the significance": Einstein (1922), reprinted in Bernstein and Feinberg (1986).

"If you find the calculations presented in my letter correct": Friedmann's letter to Einstein, 1922, in Schweber (2008), p. 324.

"there are time varying solutions": Einstein (1923), reprinted in Bernstein and Feinberg (1986).

36 "a very brilliant student": Douglas (1967).

37 H. Weyl and A. Eddington's discussions of the de Sitter effect: Weyl (1923) and Eddington (1963).

38 Vesto Slipher: The relevant papers are Slipher (1913), Slipher (1914), and Slipher (1917), which can be found at http://www.roe.ac.uk/~jap/slipher.

39 K. Lundmark's attempt at detecting the de Sitter effect: Lundmark (1924). obscure Belgian publication: Lemaître (1927).

40 "Although your calculations are correct": Einstein to G. Lemaître at the 1927 Solvay Conference, in Berger (1984).
    E. Hubble's papers measuring the distance to Andromeda: Hubble (1926) and Hubble (1929a).

41 Hubble and Humason: A fascinating description of working with E. Hubble at Palomar can be found in M. Humason's AIP interview, in Shapiro (1965).

42 E. Hubble's and M. Humason's back-to-back papers: Humason (1929) and Hubble (1929b).

43 "I send you a few copies of the paper": Letter from G. Lemaître to A. Eddington, 1930, reproduced in Nussbaumer and Bieri (2009), p. 123.

44 "If the world has begun with a single": Lemaître (1931).
    "The notion of a beginning of the present order": Eddington (1931).

46 "serious expressions on their faces": *Los Angeles Times,* January 11, 1933.
    "This is the most beautiful and satisfactory": A. Einstein about G. Lemaître in Kragh (1996), p. 55.
    "World's Leading Cosmologist": *New York Times,* February 19, 1933.

4. COLLAPSING STARS

There are a number of histories of quantum physics. I would pick Kumar (2009) as an excellent up-to-date description of the characters and concepts. The fight and fallout between Eddington and Chandra is beautifully described in Miller (2007) with a personal view (from Chandra) in Chandrasekhar (1983). In Thorne (1994), you can find how their battle fits into the grand narrative. I have not discussed the almost simultaneous discovery of Chandra's mass limit by E. Stoner and L. Landau, but it is worth having a look at Stoner (1929) and Landau (1932).

Oppenheimer is a truly fascinating character and there are a number of biographies. One of my favorites is the slim, almost personal description of the man in Bernstein (2004), but I have also used the authoritative Bird and Sherwin (2009). Monk (2012) came out as I was finishing this book and is also a wonderful resource.

48 "the star tends to close itself off from any communication": Oppenheimer and Snyder (1939).

"As you see, the war is kindly disposed toward me": K. Schwarzschild letter to
A. Einstein in Einstein (2012).

"Schwarzschild's bent was more practical": A. Eddington on K. Schwarzschild
in Eddington and Schwarzschild (1917).

49 "I had not expected that one could formulate the exact solution of the problem
in such a simple way": A. Einstein letter to K. Schwarzschild in Einstein (2012).

50 "When we obtain by mathematical analysis": Eddington (1959), p. 103.

51 "It would seem that the star will be in an awkward predicament": Ibid., p. 172.
"the force of gravitation would be so great": Ibid., p. 6.

52 "when we *prove* a result without understanding it": Ibid., p. 103.
"By mere exposure to ultraviolet light": Lenard (1906).

54 "Certainly one of the earliest motives that I had was to show the world what an
Indian could do": S. Chandrasekhar in Weart (1977).
Chandra and Sommerfeld: Sommerfeld (1923).

57 "A star of large mass cannot pass into the white dwarf stage": Chandrasekhar
(1935a).
"a reductio ad absurdum" . . . "various accidents may intervene" . . . "I think
there should be a law of nature": Eddington (1935b).
"Now, that clearly shows that": S. Chandrasekhar on A. Eddington in
Chandrasekhar (1983).

58 "was evidently much handicapped": P. Bridgeman on J. R. Oppenheimer in
Bernstein (2004).

59 "nim nim boys": W. Pauli on J. R. Oppenheimer's group in Regis (1987).

60 "with his rabid hatred of genuine Socialism" . . . "become like Hitler and
Mussolini": Gorelik (1997).

61 "a consideration of non-static solutions must be essential": Oppenheimer and
Volkoff (1939).
"The mass would produce so much curvature": Eddington (1959), p. 6.

62 N. Bohr and J. Wheeler's paper: Bohr and Wheeler (1939).

64 "gravity becomes strong enough to hold in the radiation": Eddington (1935b).
"For my part I shall only say": S. Chandrasekhar on A. Eddington in
Chandrasekhar (1983).
A. Einstein's mistaken attempt to get rid of the Schwarzschild solution: Einstein
(1939).

5. COMPLETELY CUCKOO

The creation of, and life at, the Institute for Advanced Study in Princeton is described in some detail in Regis (1987), and Einstein and Oppenheimer's relationship and times can be found in Schweber (2008). A
fascinating and articulate description of Gödel's role in general relativity
and his interaction with Einstein is in Yourgrau (2005), and a beauti-

fully crafted novel about Gödel and Turing is Levin (2010). A wonderful graphic novel on the history of twentieth-century logic is Doxiadis and Papadimitriou (2009). If you want to understand a bit more about Einstein's failed quest for unification from a modern point of view, you should read Weinberg (2009).

For the German context of Einstein's work, and general relativity specifically, I have relied on Fölsing (1998), Wazek (2010), and Cornwell (2004). The Soviet context is far trickier, and while my starting point was Graham (1993) and Vucinich (2001), information has really begun to flow out of the Soviet archives that questions some of the Western views of what was going on during that period. I have relied heavily on my colleague Dr. Andrei Starinets and his translation of archival material of the time, but a book about Landau's times, which I eagerly await in translation, is Gorobets (2008). The stagnation of general relativity in the United States can be pieced together from Thorne (1994), DeWitt-Morette (2011), and Wheeler and Ford (1998).

70 "The ideal world is nothing else than the material world": Marx (1990).

72 The private letters to Beria: ЦХСД. ф.4. Оп.9. Д.1487. Л.5–7. Копия. CDMD (Central Depository of Modern Documents of the Russian Federation Archives) and ЦХСД. Ф. 4. Оп. 9. Д. 1487. Л. 11–11 об. Копия. CDMD (Central Depository of Modern Documents of the Russian Federation Archives).

74 "Einstein on Verge of Great Discovery": *New York Times*, November 4, 1928.
"Einstein Is Amazed at Stir Over Theory": *New York Times*, February 4, 1929.

75 "New Einstein Theory Gives a Master Key to the Universe": *New York Times*, December 27, 1949.
"Einstein Offers New Theory": *New York Times*, March 30, 1953.

77 "a wonderful piece of Earth": A. Einstein letter to Queen of Belgium, 1933, kept in the Albert Einstein Archives at the Hebrew University in Jerusalem, in Fölsing (1998), p. 679.
"just for the privilege of walking home": A. Einstein on K. Gödel in Yourgrau (2005), p. 6.

78 K. Gödel's solution: Gödel (1949).

80 "an important contribution": A. Einstein on Gödel's solution in Schilpp (1949).
"Princeton is a madhouse": J. R. Oppenheimer to his brother in Schweber (2008), p. 265.
"Oppenheimer has made no contribution": W. Pauli and A. Einstein on Oppenheimer in Schweber (2008), p. 271.
"The guest list at Oppie's": *Time* magazine, November 8, 1948.

81 "the general theory of relativity is one of the least promising": F. Dyson letter, 1948, in Schweber (2008), p. 272.

"gravitation and fundamental theory": S. Goudsmit in DeWitt-Morette (2011).

82 "persistent campaign to reverse US Military Policy": *Fortune,* May 1953, in Schweber (2009), p. 181.

"We find that Dr. Oppenheimer's continuing conduct": Bernstein (2004).

"Einstein Warns World": the *New York Post,* February 13, 1950.

83 "What ought the minority of intellectuals do against": A. Einstein in the *New York Times,* June 12, 1953.

"Einstein was a physicist, a natural philosopher": J. R. Oppenheimer lecture, 1965, in Schweber (2008), p. 277.

"in the close-knit fraternity": in *Time* magazine, November 8, 1948.

84 "During the end of his life": J. R. Oppenheimer in *L'Express,* December 20, 1965.

6. RADIO DAYS

Radio astronomy and how it ended up fueling general relativity is well told in Munns (2012) and in Thorne (1994). Hoyle is a larger-than-life character, and it is definitely worth reading his autobiography, Hoyle (1994), but also the two substantial biographies, Gregory (2005) and Minton (2011). The AIP interview with Gold, Weart (1978), is very enlightening, and Kragh (1996) does an exhaustive job of mapping out the conflict with Ryle. I highly recommend reading Jansky (1933) and Reber (1940) to see how a field is discovered.

85 "These theories were based on the hypothesis": F. Hoyle in BBC Radio broadcast, 1949.

86 "a feeling that he had gone far": R. Williamson on F. Hoyle on the Canadian Broadcasting Corporation, 1951, in Kragh (1996), p. 194.

Eddington's theory: A. Eddington's fundamental theory is laid out in gory detail in Eddington (1953).

"Whether or not it will survive": E. A. Milne on Eddington's fundamental theory in Kilmister (1994), p. 3.

87 "complete nonsense: more precisely": W. Pauli on A. Eddington in Miller (2007), p. 89.

"I was allowed to drift": Lightman and Brawer (1990), p. 53.

88 "I wanted to live for the rest of my days": H. Bondi in Kragh (1996), p. 166.

"would continue . . . sometimes being rather repetitious": T. Gold in Kragh (1996), p. 186.

89 "I am afraid all we can do is to accept the paradox": W. de Sitter in Kragh (1996), p. 74.

"It was an irrational process that cannot be described": Hoyle (1950).

90 "a distinctly unsatisfactory notion": Ibid.

Dead of Night: This is a British film by Alberto Cavalcanti (1945).

91 "about one atom every century": Hoyle (1955), p. 290.

two papers: The two first steady-state papers are Bondi and Gold (1948) and Hoyle (1948).

"I do not believe the hypothesis": E. A. Milne in Kragh (1996), p. 190.

"for if there is any law which has withstood": Born (1949).

"romantic speculation": Michelmore (1962), p. 253.

"worn out with explaining points of physics": F. Hoyle in Kragh (1996), p. 192.

92 "I found it difficult to get my papers published": Ibid.

"I do not think it unreasonable to say": Ibid., p. 270.

93 The birth of radio astronomy: Jansky (1933), Reber (1940), and Reber (1944).

95 "I think the theoreticians have misunderstood": M. Ryle at the RAS, 1955, in Lang and Gingrich (1979).

96 "If we accept the conclusion that most of the radio stars": Ryle (1955).

"Don't trust them": T. Gold in Weart (1978).

97 "catalogue is compared . . . the Cambridge catalogue is affected by the low": Mills and Slee (1956).

"Radio astronomers must make considerable progress": Hanbury-Brown (1959).

"this has happened more than once": Bondi (1960), p. 167.

98 "appear to provide conclusive evidence": Ryle and Clarke (1961).

"the Bible was right": *Evening News and Star,* February 10, 1961.

"I certainly don't consider this the death": H. Bondi in the *New York Times,* February 11, 1961.

7. WHEELERISMS

Wheeler is a great character and the driving force behind modern general relativity. His biography, Wheeler and Ford (1998), candidly exposes his two sides: the "radical" and the "conservative." But, as importantly, the atmosphere at the time and the bizarre alliance between industry and relativists is well described in DeWitt and Rickles (2011) and DeWitt-Morette (2011) as well as in Mooallem (2007) and Kaiser (2000). It is worthwhile to browse through the Gravity Research Foundation website, at http://www.gravityresearchfoundation.org, where you can find DeWitt's winning essay.

The realization that quasars are cosmological is well described in Thorne (1994) and in Schmidt's interview for the AIP, Wright (1975). The atmosphere in Schild's group at Austin is described to great effect in Melia (2009), and a great firsthand account of what happened at the first Texas Symposium can be found in Schucking (1989) and Chiu (1964).

100 "my first step": Wheeler (1998), p. 228.

101 "radical conservative": A. Komar in Misner (2010).

"liked to tell us in class": Wheeler (1998), p. 87.

102 Feynman: A fascinating description of Richard Feynman's science can be found in Krauss (2012).

"by pushing a theory to its extremes": Wheeler (1998), p. 232.

104 "For many years this idea of collapse": Ibid., p. 294.

105 "space traveller": B. DeWitt's essay "Why Physics?" in DeWitt-Morette (2011).

"a sojourn [that] did not make good professional sense": S. Weinberg obituary of B. DeWitt in DeWitt-Morette (2011).

"What goes up will come down": R. Babson in GRF website.

"she was unable to fight gravity": Ibid.

106 "Space Ship Marvel Seen": *New York Herald Tribune*, November 21, 1955.

"New Air-Dream Planes": *New York Herald Tribune*, November 22, 1955.

"Future Planes": *Miami Herald*, December 2, 1955.

107 "Conquest of Gravity": *New York Herald Tribune*, November 20, 1955.

"eventually be controlled like light and radio waves": Ibid.

"grossly practical things . . . any frontal attack": B. DeWitt's winning essay, 1953, at the GRF website.

108 "gravitation has received . . . peculiarly difficult . . . fundamental equations are almost hopeless of solution . . . the phenomenon of gravitation": Ibid.

"the quickest $1000 I ever earned": B. DeWitt in DeWitt-Morette (2011).

"In the minds of the public": A. Bahnson in DeWitt and Rickles (2011).

109 "The main meeting began yesterday": Feynman (1985).

110 "There exists . . . one serious difficulty": R. Feynman in DeWitt and Rickles (2011).

"the best viewpoint": Ibid.

111 "Relativity seems almost to be a purely": R. Dicke in DeWitt and Rickles (2011).

113 "Something terrible happened at the office today": M. Schmidt in Wright (1975).

"mere peanuts by cosmological standards": *Time* magazine, November 3, 1966.

114 "American scientists outside of geophysics ": Schucking (1989).

115 "science starved south": Ibid.

"energies which lead to the formation of radio": Robinson, Schild, and Schucking (1965).

116 "incredibly beautiful": *Life* magazine, January 24, 1964.

"quasars": Chiu (1964).

"the issue of the final state": J. Wheeler in Harrison, Thorne, Wakano, and Wheeler (1965).

117 "utter disbelief . . . distinguished participant": Schucking (1989).

"The scientists, having stretched their imagination": *Life* magazine, January 24, 1964.

"Here we have a case": Robinson, Schild, and Schucking (1965).

"Let us all hope that it is right": Ibid.

8. SINGULARITIES

By far the best book on the Golden Age of General Relativity is Thorne (1994); it is exhaustive, detailed, and filled with personal anecdotes. It lays out the three main schools (Cambridge, Moscow, and Princeton) that fueled the renaissance of the field. Melia (2009) has a complementary view, describing how black hole astrophysics has developed until today. For the Soviet side of the story, there is an idiosyncratic collection of anecdotes and reminiscences about Zel'dovich and his disciples in Sunyaev (2005), some of which are developed in Novikov (2001). The discovery of pulsars is beautifully told in Bell Burnell (2004).

118 "Wheeler's talk made a real impression on me": R. Penrose, private communication, 2011.
"Golden Age of General Relativity": Thorne (1994).
119 "Well, you can ask Dennis": Ibid.
120 "support the 'old Einstein' against the new": Ibid.
"We didn't really ask where the money": Ibid.
Kerr and Penrose: A vivid description of R. Kerr and R. Penrose at the first Texas Symposium can be found in Schucking (1989).
121 "They didn't pay much attention to him": R. Penrose, private communication, 2011.
122 "Landau's Theoretical Minimum": A description can be found in Ioffe (2002).
123 "that bitch": L. Landau on Y. Zel'dovich in Gorelik (1997).
"That's it. He's gone": L. Landau in Gorelik (1997).
124 "You couldn't really prove anything doing it the way they did it": R. Penrose, private communication, 2011.
125 "Deviations from spherical symmetry": Penrose (1965).
"I hid in the corner . . . too embarrassing": R. Penrose, private communication, 2011.
126 "It was really that plot that converted Dennis": M. Rees, private communication, 2011.
129 "first couple of years involved a lot of very heavy work": Bell Burnell (2004).
"When I left I could swing a sledge hammer": Ibid.
"We had begun nicknaming": Ibid.
"Unusual signals": Hewish et al. (1968).
"journalists were asking relevant questions": Bell Burnell (1977).
130 "They'd turn to me": Bell Burnell (2004).
"The Girl Who Spotted the Little Green Men": *The Sun*, March 6, 1968.
"pulsars": *Daily Telegraph*, March 5, 1968.
"I did get to go in the end": J. Bell Burnell, private communication, 2011.

Zel'dovich: A commented collection of Zel'dovich's most significant papers can be found in Ostriker (1993).

"It is difficult, but interesting": Sunyaev (2005).

"The Godfather of psychoanalysis": Ostriker (1993).

132 "extremely massive objects of relatively small size": Salpeter (1964).

"having to drain": R. Penrose in John (1973).

"completely collapsed gravitational object": Wheeler (1998), p. 296.

"after you get around to saying": J. Wheeler in the *New York Times,* October 20, 1992.

"We would be wrong to conclude": Lynden-Bell (1969).

134 "The story of the phenomenal transformation": DeWitt and DeWitt (1973).

135 "There were three groups": M. Rees, private communication, 2011.

"Despite our desperate efforts": Novikov (2001).

"I saw black holes change": R. Penrose, private communication, 2011.

9. UNIFICATION WOES

The rise of quantum electrodynamics and the standard model has been written about in detail over the past decades. A meaty tome on the development of QED is Schweber (1994), but a much more digestible description of the history is Close (2011). DeWitt-Morette (2011) is an idiosyncratic biography of Bryce DeWitt with an interesting and varied collection of his writings. A masterful and utterly compelling biography of Dirac is Farmelo (2010), and it is worth reading some of his papers just to get a sense of the economy of prose.

The proceedings of the Oxford Symposium on Quantum Gravity in Isham, Penrose, and Sciama (1975) is fascinating, a real time capsule of what was going on at the time, but more recent reviews can be found in Duff (1993), Smolin (2000), and Rovelli (2010). A first account of the discovery of black hole radiation can be found in Hawking (1988) and Thorne (1994). Ferguson (2012) is a reasonably complete biography of Hawking, fleshing out the background to his main discovery.

137 "What is the gravitational field doing there": B. DeWitt in DeWitt-Morette (2011).

"That is a very important problem": W. Pauli to B. DeWitt in DeWitt-Morette (2011).

141 "very dissatisfied with the situation . . . This is just not sensible": Kragh (1990), page 184.

142 "Dirac was this ghost we rarely saw": G. Ellis, private communication, 2012.

144 "greeted with hoots of derision": M. Duff, private communication, 2011, and Duff (1993).

"wasn't doing physics": P. Candelas, private communication, 2011.

"What God hath torn asunder": Isham, Penrose, and Sciama (1975).

145 "It appears that the odds are stacked against us": M. Duff in Isham, Penrose, and Sciama (1975).

Nature article on the symposium: The write-up of the Oxford symposium was anonymous in *Nature*, 248, 282 (1974).

148 "We emphasize that one should not regard T": Bekenstein (1973).

"evaporate": Hawking (1974).

"a fairly small explosion": Ibid.

149 "People treated Hawking with great respect": P. Candelas, private communication, 2011.

150 "I was greeted with general incredulity": Hawking (1988).

"the main attraction": *Nature*, 248, 282 (1974).

"one of the most beautiful": D. Sciama in Boslough (1989).

"like candy rolling on the tongue": J. Wheeler as reported by B. Carr in *The Observer*, January 1, 2012.

10. SEEING GRAVITY

The tragic story of Joseph Weber is well known in the field but not often written about. Collins (2004) is a thorough study of the development of gravitational wave physics by a sociologist. He started interviewing the participants when Weber was still on a high, and his book is full of interviews and quotes. It is a must-read if you want to get the full story of how the field has developed and the battles that proponents of LIGO had to fight in order to build it. Thorne (1994) is an insider's view of the story by the elder statesman of gravitational wave physics. Kennefick (2007) does an excellent job of discussing the roots of the field and filling in the back story, and Bartusiak (1989) and the more up-to-date Gibbs (2002) summarize progress at different stages. The history of numerical relativity is neatly summarized in Appell (2011).

It is well worth looking at some of the original material. For example, the discussion of the reality of gravitational waves at the Chapel Hill meeting in DeWitt and Rickles (2011) is fascinating. Weber's sequence of papers — Weber (1969), Weber (1970a), Weber (1970b), and Weber (1972) — is a march toward greater certainty. He is then brutally struck down in Garwin (1974).

152 "We're number one in the field": J. Weber in the *Baltimore Sun*, April 7, 1991.

154 "speed of thought": A. Eddington in Kennefick (2007).

reality of gravitational waves: A discussion of the reality of gravitational waves can be found in DeWitt and Rickles (2011).

157 "A good feature is the fact": Weber (1970b).

Weber's results: Coverage of Weber's results can be found in *Time* magazine and the *New York Times* in 1970.

sources of gravitational radiation: A review of the hypothetical sources of gravitational radiation at the time can be found in Tyson and Giffard (1978).

158 "Since the high rate of mass loss": Sciama, Field, and Rees (1969).

159 "people were very suspicious": B. Schutz, private communication, 2012.

160 "*did not* result from gravity waves": Garwin (1974).

161 Taylor's results: Taylor's plot was shown at the ninth Texas Symposium in Munich, 1978, and the proceedings were published as Ehlers, Perry, and Walker (1980).

164 "either the programmer will shoot himself": C. Misner in DeWitt and Rickles (2011).

165 solving for colliding black holes on a computer: The first steps are described by L. Smarr in Christensen (1984).

"Naive things weren't working": F. Pretorius, private communication, 2011.

166 "There was a serious possibility": Ibid.

168 "Most of the astrophysical community": A. Tyson in the *New York Times*, April 30, 1991.

"should wait for someone to come up": J. Ostriker in the *New York Times*, April 30, 1991.

169 "under the radar": F. Pretorius, private communication, 2011.

"pure agony": Ibid.

"There was quite a bit of excitement": Ibid.

170 "A great man is not afraid to admit publicly": F. Dyson in Collins (2004).

171 "by the time he was opposing LIGO": B. Schutz, private communication, 2012.

11. THE DARK UNIVERSE

The phenomenal success story of modern cosmology is well documented. Peebles, Page, and Partridge (2009) includes a list of testimonials and essays with a description of the rise and rise of the field. It is well worth reading some of the books that cropped up along the way, such as Overbye (1991) or the compilation of interviews in Lightman and Brawer (1990). A personal memoir of the COBE discovery is Smoot and Davidson (1995) with a more journalistic take in Lemonick (1995). Panek (2011) is a fantastic description of the march toward the cosmological constant during the late 1990s with much of the gory detail of who did what in the supernova searches. The AIP interviews with Peebles — Harwitt (1984), Lightman (1988b), and Smeenk (2002) — are a wonderful source for his view of the universe. For more detailed explanations of our current theory of the universe, you might read Silk (1989) and Ferreira (2007). It is well worth browsing through some of the main early papers of modern cosmology in Bernstein and Feinberg (1986) and taking a look at the Einstein Centenary proceedings, Hawking and Israel (1979), and the Critical Dialogues proceedings, Turok (1997).

173 "a fundamental science . . . the grandest of environmental sciences": M. Rees in Turok (1997).

175 "the cosmological constant": Peebles (1971).
"the dirty little secret": J. Peebles, private communication, 2011.

176 "Rather quickly in my career": J. Peebles in Smeenk (2002).
"a limited subject . . . a science with two or three numbers": J. Peebles in Lightman (1988b).
"To him physics was certainly theory but it had to lead": J. Peebles in Smeenk (2002).
"We've been scooped": R. Dicke as told by J. Peebles in Smeenk (2002).

177 a difficult, open problem that hardly anyone wanted to work on: While Peebles and his contemporaries really established the field of physical cosmology, the idea that there is some fundamental connection between the expanding hot Big Bang model and the formation of galaxies appears first in Lemaître (1934) and Gamow (1948).

178 large structures: The ideas leading up to the formation of large-scale structure can be found in Silk (1968), Sachs and Wolfe (1967), Peebles and Yu (1970), and Zel'dovich (1972).

179 "No one paid any attention to our paper": J. Peebles, private communication, 2011.
"stream of galaxies . . . supergalaxy": G. de Vaucouleurs in Lightman (1988a).
"We have no evidence for the existence": Ibid.

180 "Superclustering is nonexistent": Ibid.
"good observations are worth more than another mediocre theory": M. Davis on Peebles in Lightman and Brawer (1990).
"flabbergasted . . . I wrote some pretty vitriolic papers with examples": J. Peebles in Lightman (1988b).

182 "inner space and outer space": A historic conference on connecting "inner space" and "outer space" was held at Fermilab in 1984 and written up in Kolb et al. (1986).

183 "The density of luminous matter": F. Zwicky in Panek (2011), p. 48.

184 "we think it likely that the discovery of invisible matter": Faber and Gallagher (1979).
"I didn't take it at all seriously": J. Peebles, private communication, 2011.

185 "There was a lot of net casting in the eighties": J. Peebles in Smeenk (2002).

186 Y. Zel'dovich's estimate of the cosmological constant: Zel'dovich (1968).
"We argue here that the successes": Efstathiou, Sutherland, and Maddox (1990).

187 "a universe having critical energy density": Ostriker and Steinhardt (1995).

"The problem with the choice": Peebles (1984).

"A non-zero cosmological constant": Efstathiou, Sutherland, and Maddox (1990).

"requires a seemingly implausible": Blumenthal, Dekel, and Primack (1988).

"How can we explain the non-zero": Ostriker and Steinhardt (1995).

188 "If you're religious, this is like looking at God": G. Smoot press conference at Lawrence Berkeley Laboratory, 1992.

190 "The findings also appear to breathe": *Washington Post,* January 9, 1998.

"Exploding Stars Point to a": Glanz (1998).

"stunned the universe may be accelerating": CNN, February 27, 1998.

"My own reaction is somewhere between amazement and horror": B. Schmidt in the *New York Times,* March 3, 1998.

191 "The best explanation for what the data": J. Peebles, private communication, 2011.

192 "After a genie is let out of the bottle": Zel'dovich and Novikov (1971), p. 29.

dark energy: The term *dark energy* was first proposed in Huterer and Turner (1998).

12. THE END OF SPACETIME

The modern history of quantum gravity is fraught and fascinating. To get a grand overview, Rovelli (2010) has an appendix with the various major stages, discoveries, and shifts. DeWitt-Morette (2011) describes the genesis of the "Trilogy" and how DeWitt viewed the development of the field. For a hugely successful and articulate summary of string theory, you need to turn to Greene (2000). Yau and Nadis (2010) takes a mathematician's viewpoint of string theory. The alternative paths to quantum gravity, such as loop quantum gravity, are well described in Smolin (2000). The two books that led to the vicious backlash against string theory are Smolin (2006) and Woit (2007). It is worth looking at some of the blogs and following the discussions to see how heated they became. I would look at the following and wind back to when the books were published:

http://blogs.discovermagazine.com/cosmicvariance/

http://asymptotia.com/

http://www.math.columbia.edu/~woit/wordpress/

The black hole information paradox is an ongoing story, and even though I haven't discussed "black hole complementarity," I highly recommend Susskind (2008) for a personal and energetic account of how the paradox has developed over the years. Solutions are still cropping up: as I was finishing this book, another proposal, "the firewall," which modifies one of the fundamental tenets of general relativity, was being heatedly debated. For a description of the proposal, see http://blogs.scientific american.com/critical-opalescence/2012/12/14/when-you-fall-into-a-black-hole-how-long-have-you-got/.

194 "Is the End in Sight for Theoretical Physics?": S. Hawking's lecture is published in its entirety in Boslough (1989).

195 Hawking's lecture: A colorful description of Hawking's talk can be found in Susskind (2008).

197 "Trilogy": DeWitt-Morette (2011).
"Wheeler got tremendously excited": Ibid.

198 "a sort of patron of string theory . . . a conservationist . . . I set up a nature reserve": Interview with M. Gell-Mann in *Science News*, September 15, 2009.

199 "M stands for Magic": E. Witten in interview with Swedish public radio, June 6, 2008.

200 "I think all this superstring stuff is crazy . . . I don't like that they're": R. Feynman in Davies and Brown (1988), p. 194.
"superstring physicists have not yet shown": S. Glashow in Davies and Brown (1988).
"The long-standing crisis of string theory": Friedan (2002).

203 "In viewing string theory": DeWitt-Morette (2011).
"should be confined to the dustbins of history . . . violates the very spirit of relativity": Ibid.
"the Wheeler-DeWitt equation is wrong": Ibid.
"elegant . . . apart from some": Ibid.
"We have made tremendous progress with string and M-theory": M. Duff, private communication, 2011.
"M-theory is the *only* candidate": Hawking and Mlodinow (2010), p. 181.

204 "quantum gravity . . . loop quantum gravity": M. Duff, private communication, 2011.
"They can't even calculate what a graviton does": P. Candelas, private communication, 2011.
"A lot of people are frustrated that this community": L. Smolin in *Wired*, September 14, 2006.
annual string theory meeting: In 2008, at the annual jamboree for string theory — Strings 2008 held at CERN — Rovelli was finally invited to make the case for loop quantum gravity.

205 "It's a deal breaker": Episode 2, Series 2, of *The Big Bang Theory*, Chuck Lorre Productions/CBS.

207 "approximate, emergent, classical concept": Witten (1996a).
"geometry in the small": Wheeler (1955).

13. A SPECTACULAR EXTRAPOLATION

Not a lot has been written on modified theories of gravity that I can recommend. Barrow and Tipler (1988) and Barrow (2003) do an excellent job of discussing the large number problem that intrigued Dirac, which is also discussed in Farmelo (2010). Sakharov's scientific interests are cursorily discussed in Lourie (2002) and his own autobiography, Sakharov (1992). I recommend you take a peek at his collected works in Sakharov (1982) to see how concise he was. For the history of Milgrom and Bekenstein's theory, it is probably best to read one of Bekenstein's reviews; for example, Bekenstein (2007) is quite technical but will give you a flavor of what is going on. Peebles (2004) is a statesmanlike review of why looking beyond general relativity might be a good thing, and a more lay account can be found in Ferreira (2010).

212 "The beauty of the equations provided by nature": P. Dirac, interviewed on Canadian radio, 1979.
"Experimentalists, especially those at NASA . . . as time went by": Brans (2008).

213 "a man of universal interests": A. Sakharov on Y. Zel'dovich in Sakharov (1988).
"I don't understand how Sakharov thinks": Y. Zel'dovich on A. Sakharov, http://www.joshuarubenstein.com/KGB/KGB.html.

215 "People like Hawking are devoted to science": Y. Zel'dovich on A. Sakharov in Sunyaev (2005).
"I felt compelled to speak out": Sakharov (1992).

217 "The elegant logic of general relativity . . . a spectacular extrapolation": Peebles (2000).

218 "At some point, I felt": J. Bekenstein, private communication, 2011.
"Some looked at me as if I told them": Ibid.

219 "a dirty word": N. Turok, private communication, 2005.
"By no means have we ruled out MOND": J. Peebles in Smeenk (2002).
"One has to take into account": J. Bekenstein, private communication, 2011.
"I decided that it was time": Ibid.
Bekenstein's theory: Bekenstein (2004).

14. SOMETHING IS GOING TO HAPPEN

If you want to come to grips with the multiverse, you might want to try two of its most eloquent advocates, such as Susskind (2006) and Greene (2012), but temper them with the contrasting view of Ellis (2011b). If you want to follow the big experiments, you should check out websites such as the following:

http://www.skatelescope.org/

http://www.eventhorizontelescope.org/

http://www.ligo.caltech.edu/

These are full of interesting facts about what is actually going on at the coal face of observational research in general relativity.

226 textbooks: The two classic textbooks I describe are Misner, Thorne, and Wheeler (1973) and Weinberg (1972).

227 Event Horizon Telescope: A description of the Event Horizon Telescope can be found at http://www.eventhorizontelescope.org/.

228 "Black Hole Confirmed in Milky Way": http://news.bbc.co.uk/2/hi/science/nature/7774287.stm.

"Evidence Points to Black Hole": *New York Times,* September 6, 2001.

recent observation of a quasar: M. Capellari is asked about the biggest black hole discovered to date at http://www.bbc.co.uk/news/science-environment-16034045.

229 black holes at the LHC: An entertaining example of a response against black holes in the LHC can be found at http://www.lhcdefense.org/press.php.

"That is like asking what is north of the North Pole": Jocelyn Bell Burnell, private communication, 2011.

231 "I do not believe the existence": Ellis (2011b).

"The multiverse argument is a well-founded": Ellis (2011a).

"I hope the current": E. Witten in Battersby (2005).

Bibliography

BOOKS

Barrow, J., *The Constants of Nature,* Vintage (2003).

Barrow, J., P. Davies, and C. Harper Jr., *Science and Ultimate Reality: Quantum Theory, Cosmology and Complexity,* Cambridge University Press (2004).

Barrow, J., and F. Tipler, *The Anthropic Cosmological Principle,* Oxford University Press (1988).

Baum, R., and W. Sheehan, *In Search of the Planet Vulcan: The Ghost in Newton's Clockwork Universe,* Basic Books (1997).

Berendzen, R., R. Hart, and D. Seeley, *Man Discovers the Galaxies,* Science History Publications (1976).

Berger, A., *The Big Bang and Georges Lemaître,* D. Reidel (1984).

Bernstein, J., *Oppenheimer: Portrait of an Enigma,* Ivan R. Dee (2004).

Bernstein, J., and G. Feinberg, *Cosmological Constants: Papers in Modern Cosmology,* Columbia University Press (1986).

Bird, K., and M. Sherwin, *American Prometheus: The Triumph and Tragedy of J. Robert Oppenheimer,* Atlantic (2009).

Bodanis, D., *E=mc²: A Biography of the World's Most Famous Equation,* Pan (2001).

——, *Electric Universe: How Electricity Switched On the Modern World,* Abacus (2006).

Bondi, H., *Cosmology,* Cambridge University Press (1960).

Boslough, J., *Stephen Hawking's Universe,* Avon (1989).

Burbidge, G., and M. Burbidge, *Quasi-Stellar Objects,* W. H. Freeman (1967).

Chandrasekhar, S., *Eddington: The Most Distinguished Astrophysicist of His Time,* Cambridge University Press (1983).

Christensen, S., ed., *Quantum Theory of Gravity: Essays in Honor of the 60th Birthday of Bryce S. DeWitt,* Adam Hilger (1984).

Close, F., *The Infinity Puzzle,* Oxford University Press (2011).

Collins, H., *Gravity's Shadow: The Search for Gravitational Waves,* University of Chicago Press (2004).

Cook, N., *The Hunt for Zero Point,* Arrow (2001).

Cornwell, J., *Hitler's Scientists: Science, War, and the Devil's Pact,* Penguin (2004).

Danielson, D., *The Book of the Cosmos: Imagining the Universe From Heraclitus to Hawking*, Perseus (2000).

Davies, P., and J. Brown, eds., *Superstrings*, Cambridge University Press (1988).

DeWitt, C., and B. DeWitt, eds., *Relativity Groups and Topology*, Gordon and Breach Science Publishers (1964).

——, eds., *Black Holes*, Gordon and Breach Science Publishers (1973).

DeWitt, C., and D. Rickles, *The Role of Gravitation in Physics: Report from the 1957 Chapel Hill Conference*, Edition Open Access (2011).

DeWitt-Morette, C., *Gravitational Radiation and Gravitational Collapse*, D. Reidel (1974).

——, *The Pursuit of Quantum Gravity: Memoirs of Bryce DeWitt From 1946 to 2004*, Springer (2011).

Dickens, C., *A Detective Police Party*, Read Books (2011).

Doxiadis, A., and C. Papadimitriou, *Logicomix: An Epic Search for Truth*, Bloomsbury (2009).

Durham, F., and R. Purrington, *Frame of the Universe: A History of Physical Cosmology*, Columbia University Press (1983).

Eddington, A., *The Nature of the Physical World*, Cambridge University Press (1929).

——, *Fundamental Theory*, Cambridge University Press (1953).

——, *The Internal Constitution of the Stars*, Dover (1959).

——, *The Mathematical Theory of Relativity*, Cambridge University Press (1963).

Ehlers, J., J. Perry, and M. Walker, *9th Texas Symposium on Relativistic Astrophysics*, New York Academy of Sciences (1980).

Einstein, A., *Relativity*, Routledge Classics (2001).

——, *The Collected Papers of Albert Einstein*, Volumes 1–13, Princeton University Press (2012).

Eisenstaedt, J., *The Curious History of Relativity: How Einstein's Theory of Gravity Was Lost and Found Again*, Princeton University Press (2006).

Eisenstaedt, J., and A. Kox, eds., *Studies in the History of General Relativity*, Volume 3, Birkhauser (1992).

Ellis, G., A. Lanza, and J. Miller, *The Renaissance of General Relativity and Cosmology*, Cambridge University Press (1993).

Farmelo, G., *The Strangest Man: The Life of Paul Dirac*, Faber and Faber (2010).

Ferguson, K., *Stephen Hawking: His Life and Work*, Bantam (2012).

Ferreira, P., *The State of the Universe: A Primer in Modern Cosmology*, Phoenix (2007).

Feynman, R., *Surely You're Joking, Mr. Feynman! Adventures of a Curious Character*, W. W. Norton (1985).

Feynman, R., F. Morinigo, and W. Wagner, *Lectures on Gravitation*, Penguin (1999).

Fölsing, A., *Albert Einstein*, Penguin (1998).

Gamow, G., *My World Line: An Informal Autobiography*, Viking (1970).

Gorobets, B., *The Landau Circle: The Life of a Genius*, URSS (2008).

Graham, L., *Science in Russia and in the Soviet Union: A Short History*, Cambridge University Press (1993).

Greene, B., *The Elegant Universe: Superstrings, Hidden Dimensions, and the Quest for the Ultimate Theory*, Vintage (2000).

——, *The Hidden Reality: Parallel Universes and the Deep Laws of the Cosmos*, Penguin (2012).

Gregory, J., *Fred Hoyle's Universe*, Oxford University Press (2005).

Gribbin, J., and M. Gribbin, *How Far Is Up: The Men Who Measured the Universe*, Icon Books (2003).

Harrison, B., K. Thorne, M. Wakano, and J. Wheeler, *Gravitation Theory and Gravitational Collapse*, University of Chicago Press (1965).

Harvey, A., *On Einstein's Path: Essays in Honor of Engelbert Schucking*, Springer-Verlag (1992).

Hawking, S., *A Brief History of Time: From the Big Bang to Black Holes*, Bantam (1988).

Hawking, S., and W. Israel, eds., *General Relativity: An Einstein Centenary Survey*, Cambridge University Press (1979).

——, eds., *Three Hundred Years of Gravitation*, Cambridge University Press (1989).

Hawking, S., and L. Mlodinow, *The Grand Design*, Random House (2010).

Hoyle, F., *The Nature of the Universe*, Oxford Blackwell (1950).

——, *Frontiers of Astronomy*, Mentor (1955).

——, *Home Is Where the Wind Blows: Chapters From a Cosmologist's Life*, University Science Books (1994).

Hoyle, F., G. Burbidge, and J. Narlikar, *A Different Approach to Cosmology: From a Static Universe Through the Big Bang Towards Reality*, Cambridge University Press (2000).

Isaacson, W., *Einstein: His Life and Universe*, Pocket Books (2008).

Isham, C., R. Penrose, and D. Sciama, eds., *Quantum Gravity: An Oxford Symposium*, Clarendon (1975).

John, L., *Cosmology Now*, BBC (1973).

Kaiser, D., "Making Theory: Producing Physics and Physicists in Postwar America," unpublished PhD thesis, Harvard University (2000).

Kennefick, D., *Traveling at the Speed of Thought: Einstein and the Quest for Gravitational Waves*, Princeton University Press (2007).

Kilmister, C., *Eddington's Search for a Fundamental Theory: A Key to the Universe*, Cambridge University Press (1994).

Kolb, E., M. Turner, K. Olive, and D. Seckel, *Inner Space/Outer Space*, University of Chicago Press (1986).

Kragh, H., *Dirac: A Scientific Biography*, Cambridge University Press (1990).

——, *Cosmology and Controversy: The Historical Development of Two Theories of the Universe*, Princeton University Press (1996).

Krauss, L., *Quantum Man: Richard Feynman's Life in Science*, W. W. Norton (2012).

Kumar, M., *Quantum: Einstein, Bohr, and the Great Debate About the Nature of Reality*, Icon (2009).

Lambert, D., *Un atome d'univers: La vie et l'oeuvre de Georges Lemaître*, Éditions Racine (1999).

Lang, K., and O. Gingrich, *A Source Book in Astronomy and Astrophysics, 1900–1975*, Harvard University Press (1979).

Lemonick, M., *The Light at the Edge of the Universe*, Princeton University Press (1995).

Lenin, V., *Materialism and Empiriocriticism*, Literary Licensing, LLC (2011).

Levin, J., *A Madman Dreams of Turing Machines*, Phoenix (2010).

Lichnerowicz, A., A. Mercier, and M. Kervaire, *Cinquantenaire de la théorie de la relativité*, Birkhäuser (1956).

Lightman, A., and R. Brawer, *Origins: The Lives and Worlds of Modern Cosmologists*, Harvard University Press (1990).

Lourie, R., *Sakharov: A Biography*, Brandeis (2002).

Marx, K., *Capital*, Penguin (1990).

Melia, F., *Cracking the Einstein Code: Relativity and the Birth of Black Hole Physics*, University of Chicago Press (2009).

Michelmore, P., *Einstein: Profile of the Man*, Dodd, Mead (1962).

Miller, A., *Empire of the Stars: Friendship, Obsession, and Betrayal in the Quest for Black Holes*, Abacus (2007).

——, *Deciphering the Cosmic Number: The Strange Friendship of Wolfgang Pauli and Carl Jung*, W. W. Norton (2009).

Minton, S., *Fred Hoyle: A Life in Science*, Cambridge University Press (2011).

Misner, C., K. Thorne, and J. Wheeler, *Gravitation*, W. H. Freeman (1973).

Monk, R., *Inside the Centre: The Life of J. Robert Oppenheimer*, Jonathan Cape (2012).

Munns, D., *A Single Sky: How an International Community Forged the Science of Radio Astronomy*, MIT Press (2012).

North, J., *The Measure of the Universe: A History of Modern Cosmology*, Dover (1965).

Novikov, I., *River of Time*, Cambridge University Press (2001).

Nussbaumer, H., and L. Bieri, *Discovering the Expanding Universe*, Cambridge University Press (2009).

Ostriker, J., *Selected Works of Yakov Borisovich Zeldovich*, Princeton University Press (1993).

Overbye, D., *Lonely Hearts of the Cosmos*, Harper Collins (1991).

Pais, A., *Subtle Is the Lord: The Science and Life of Albert Einstein*, Oxford University Press (1982).

Pais, A., and R. Crease, *J. Oppenheimer: A Life*, Oxford University Press (2006).

Panek, R., *The 4% Universe: Dark Matter, Dark Energy, and the Race to Discover the Rest of Reality*, Houghton Mifflin Harcourt (2011).

Peat, D., *Superstrings and the Search for the Theory of Everything*, Contemporary Books (1988).

Peebles, P., *Physical Cosmology*, Princeton University Press (1971).

Peebles, P., L. Page, and B. Partridge, *Finding the Big Bang*, Cambridge University Press (2009).

Proust, M., *In Search of Lost Time*, Volume 5: *The Captive and the Fugitive*, Vintage (1996).

Regis, E., *Who Got Einstein's Office? Eccentricity and Genius at the Princeton Institute for Advanced Study*, Penguin (1987).

Reid, C., *Hilbert*, Springer-Verlag (1970).

Robinson, I., A. Schild, and E. Schucking, *Quasi-stellar Sources and Gravitational Collapse*, University of Chicago Press (1965).

Rovelli, C., *Quantum Gravity*, Cambridge University Press (2010).

Sakharov, A., *Collected Scientific Works*, Marcel Dekker (1982).

——, *Memoirs*, Vintage (1992).

Schilpp, P., *Albert Einstein: Philosopher-Scientist*, Open Court (1949).

Schrödinger, E., *Space-Time Structure*, Cambridge University Press (1960).

Schweber, S., *QED and the Men Who Made It*, Princeton University Press (1994).

——, *Einstein and Oppenheimer: The Meaning of Genius*, Harvard University Press (2008).

Silk, J., *The Big Bang*, W. H. Freeman (1989).

Smolin, L., *Three Roads to Quantum Gravity*, Weidenfeld & Nicholson (2000).

——, *The Trouble with Physics: The Rise of String Theory, the Fall of Science, and What Comes Next*, Allen Lane (2006).

Smoot, G., and K. Davidson, *Wrinkles in Time: The Imprint of Creation*, Abacus (1995).

Sommerfeld, A., *Atomic Structure and Spectral Lines*, Methuen (1923).

Stachel, J., ed., *Einstein's Miraculous Year: Five Papers That Changed the Face of Physics*, Princeton University Press (1998).

Stalin, J., *Problems of Leninism*, Foreign Languages Press (1976).

Stanley, M., *Practical Mystic*, University of Chicago Press (2007).

Sunyaev, R., ed., *Zeldovich: Reminiscences*, Taylor & Francis (2005).

Susskind, L., *The Cosmic Landscape: String Theory and the Illusion of Intelligent Design,* Back Bay Books (2006).

—— , *The Black Hole War: My Battle With Stephen Hawking to Make the World Safe for Quantum Mechanics,* Back Bay Books (2008).

Thorne, K., *Black Holes and Time Warps: Einstein's Outrageous Legacy,* Picador (1994).

Tropp, E., V. Frenkel, and A. Chernin, *Alexander A. Friedmann: The Man Who Made the Universe Expand,* Cambridge University Press (1993).

Turok, N., ed., *Critical Dialogues in Cosmology,* World Scientific (1997).

Vucinich, A., *Einstein and Soviet Ideology,* Stanford University Press (2001).

Wazek, M., *Einsteins Gegner,* Campus Verlag (2010).

Weinberg, S., *Gravitation and Cosmology,* John Wiley and Sons (1972).

—— , *Lake Views: This World and the Universe,* Harvard University Press (2009).

Wheeler, J., *Geometrodynamics,* Academic Press (1962).

—— , *At Home in the Universe,* AIP Press (1994).

Wheeler, J., and K. Ford, *Geons, Black Holes, and Quantum Foam: A Life in Physics,* W. W. Norton (1998).

Woit, P., *Not Even Wrong: The Failure of String Theory and the Continuing Challenge to Unify the Laws of Physics,* Vintage (2007).

Yau, S-T., and S. Nadis, *The Shape of Inner Space: String Theory and the Geometry of the Universe's Hidden Dimensions,* Basic Books (2010).

Yourgrau, P., *A World Without Time: The Forgotten Legacy of Gödel and Einstein,* Allen Lane (2005).

Zel'dovich, Y., and I. Novikov, *Relativistic Astrophysics: Stars and Relativity,* University of Chicago Press (1971).

ARTICLES

Abadies, J., http://arxiv.org/abs/1003.2480 (2010).

Abramowicz, M., and P. Fragile, http://arxiv.org/abs/1104.5499 (2011).

Albrecht, A., and P. Steinhardt, *Phys. Rev. Lett.,* 48, 1220 (1982).

Alpher, R., H. Bethe, and G. Gamow, *Nature,* 73, 803 (1948).

Altshuler, B., http://arxiv.org/abs/hep-ph/0207093 (2002).

Appell, D., *Physics World,* October, 36 (2011).

Ashtekhar, A., *Phys. Rev. Lett.,* 57, 2244 (1986).

—— , *Phys. Rev. D,* 36, 1587 (1987).

Ashtekhar, A., and R. Geroch, *Rep. Prog. Phys.,* 37, 122 (1974).

Bahcall, N., et al., *Science,* 284, 1481 (1999).

Barbour, J., *Nature,* 249, 328 (1974).

Barreira, M., M. Carfora, and C. Rovelli, http://arxiv.org/abs/gr-qc/9603064 (1996).

Bartusiak, M., *Discovery,* August, 62 (1989).

Battersby, S., *New Scientist,* April, 30 (2005).

Bekenstein, J., *Phys. Rev. D,* 7, 2333 (1973).

—— , *Phys. Rev. D,* 11, 2072 (1975).

—— , *Sci. Am.,* August, 58 (2003).

—— , http://arxiv.org/abs/astro-ph/0403694 (2004).

—— , http://arxiv.org/abs/astro-ph/0701848 (2007).

Bekenstein, J., and A. Meisels, *Phys. Rev. D,* 18, 4378 (1978).

—— , *Phys. Rev. D,* 22, 1313 (1980).

Bekenstein, J., and M. Milgrom, *Astroph. Jour.*, 286, 7 (1984).

Belinsky, V., I. Khalatnikov, and E. Lifshitz, *Advances in Physics*, 19, 525 (1970).

Bell Burnell, J., *Ann. New York Ac. Sci.*, 302, 665 (1977).

———, *Astron. & Geoph.*, 47, 1.7 (2004).

Blandford, R., and M. Rees, *Mon. Not. Roy. Ast. Soc.*, 169, 395 (1974).

Blumenthal, G., A. Dekel, and J. Primack, *Astroph. Jour.*, 326, 539 (1988).

Bohr, N., and J. Wheeler, *Phys. Rev.*, 56, 426 (1939).

Bondi, H., and T. Gold, *Mon. Not. Roy. Ast. Soc.*, 108, 252 (1948).

Born, M., *Nature*, 164, 637 (1949).

Bowden, M., *Atlantic Monthly*, July (2012).

Brans, C., http://arxiv.org/abs/gr-qc/0506063 (2005).

———, *AIP Conf. Proc.*, 1083, 34 (2008).

Calder, L., and O. Lahav, *Astron. & Geoph.*, 49, 1.13 (2008).

Candelas, P., et al., *Nuc. Phys. B*, 258, 46 (1985).

Carroll, S., W. Press, and E. Turner, *Ann. Rev. Astron. Astroph.*, 30, 499 (1992).

Carter, B., *Phys. Rev.*, 141, 1242 (1966).

———, *Phys. Rev.*, 174, 1559 (1968).

———, http://arxiv.org/abs/gr-qc/0604064 (2006).

Centrella, J., et al., *Rev. Mod. Phys.*, 82, 3069 (2010).

Chandrasekhar, S., *Astroph. Journ.*, 74, 81 (1931a).

———, *Mon. Not. Roy. Ast. Soc.*, 91, 456 (1931b).

———, *The Observatory*, 57, 373 (1934).

———, *The Observatory*, 58, 33 (1935a).

———, *Mon. Not. Roy. Ast. Soc.*, 95, 207 (1935b).

———, *Mon. Not. Roy. Ast. Soc.*, 95, 226 (1935c).

Chandrasekhar, S., and C. Miller, *Mon. Not. Roy. Ast. Soc.*, 95, 673 (1935).

Chandrasekhar, S., and J. Wright, *Proc. Nat. Ac. Sci.*, 47, 341 (1961).

Chiu, H., *Physics Today*, May, 21 (1964).

Choptuik, M., *Astron. Soc. Pac.*, 123, 305 (1997).

Coles, P., http://arxiv.org/abs/astro-ph/0102462 (2001).

Crease, R., *Physics World*, January, 19 (2010).

Davis, M., et al., *Astroph. Jour.*, 292, 371 (1985).

———, *Nature*, 356, 489 (1992).

de Bernardis, P., et al., *Nature*, 404, 955 (2000).

de Sitter, W., *Proc. Roy. Neth. Ac. Art. Sci.*, 20, 229 (1918).

———, *The Observatory*, 53, 37 (1930).

DeVorkin, D., interview with V. Rubin for AIP, http://www.aip.org/history/ohilist/5920_1.html (1984).

DeWitt, B., *Phys. Rev.*, 160, 1113 (1967a).

———, *Phys. Rev.*, 162, 1195 (1967b).

———, *Phys. Rev.*, 162, 1239 (1967c).

———, *Gen. Rel. Grav.*, 41, 413 (2009).

Dicke, R., et al., *Astroph. Jour.*, 142, 414 (1965).

Dirac, P., *Nature*, 168, 906 (1958a).

———, *Proc. Roy. Soc. Lon. A*, 246, 333 (1958b).

———, *Proc. Roy. Soc. A.*, 338, 439 (1974).

Doroshkevich, A., R. Sunyaev, and Y. Zel'dovich, *IAU Symp.*, 63, 213 (1974).

Doroshkevich, A., Y. Zel'dovich, and I. Novikov, *Sov. Ast.*, 11, 233 (1967).

Douglas, D., *Jour. Roy. Ast. Soc. Can.*, 61, 77 (1967).

Duff, M., *Phys. Rev. D*, 7, 2317 (1971).

———, *New Scientist*, January, 96 (1977).

———, http://arxiv.org/abs/hep-th/9308075 (1993).

———, *Sci. Am.*, February, 64 (1998).

———, http://arxiv.org/abs/1112.0788 (2011).

Dyson, F., A. Eddington, and C. Davison, *Phil. Trans. Roy. Soc. Lon.*, A 220, 291 (1920).

Earman, J., and C. Glymour, *Arch. Hist. Exac. Sci.*, 19, 291 (1978).

Eddington, A., *The Observatory*, 36, 62 (1913).

———, *The Observatory*, 38, 93 (1915).

———, *The Observatory*, 39, 270 (1916).

———, *The Observatory*, 40, 93 (1917).

———, *The Observatory*, 42, 119 (1919a).

———, *Nature*, 114, 372 (1919b).

———, *Proc. Roy. Soc. Lon. A*, 102, 268 (1922).

———, *Mon. Not. Roy. Ast. Soc.*, 90, 668 (1930).

———, *Nature*, 127, 447 (1931).

———, *Mon. Not. Roy. Ast. Soc.*, 95, 194 (1935a).

———, *The Observatory*, 58, 33 (1935b).

———, *Mon. Not. Roy. Ast. Soc.*, 96, 20 (1935c).

———, *Proc. Roy. Soc. Lon. A*, 162, 55 (1937).

———, *Proc. Phys.*, 54, 491 (1942).

———, *The Observatory*, 37, 5 (1943).

———, *Mon. Not. Roy. Ast. Soc.*, 104, 20 (1944).

Eddington, A., and K. Schwarzschild, *Mon. Not. Roy. Ast. Soc.*, 77, 314 (1917).

Efstathiou, G., W. Sutherland, and S. Maddox, *Nature*, 348, 705 (1990).

Einstein, A., *Ann. Phys.*, 17, 891 (1905a).

———, *Ann. Phys.*, 18, 639 (1905b).

———, *Ann. Phys.*, 19, 289 (1906a).

———, *Ann. Phys.*, 19, 371 (1906b).

———, *Jahr. Rad. Elek.*, 4, 411 (1907).

———, *Ann. Phys.*, 35, 989 (1911).

———, *Sitzungsberichte de Preussischen Akad. d. Wiss.*, 315 (1915).

———, *Sitzungsberichte de Preussischen Akad. d. Wiss.*, 142 (1917).

———, *Zeitschrift für Physik*, 11, 326 (1922).

———, *Zeitschrift für Physik*, 16, 228 (1923).

———, *Philosophy of Science*, 1, 163 (1934).

———, *Ann. Math.*, 40, 992 (1939).

———, *Physics Today*, August, 45 (1982).

Einstein, A., and M. Grossman, *Zeitschrift für Physik*, 62, 225 (1913).

Ellis, G., http://www.st-edmunds.cam.ac.uk/faraday/cis/Ellis (2007).

———, *Nature*, 469, 294 (2011a).

———, *Sci. Am.*, August, 38 (2011b).

Esposito, G., http://arxiv.org/abs/1108.3269v1 (2011).

Faber, S., and J. Gallagher, *Ann. Rev. Astron. Astroph. I*, 17, 135 (1979).

Ferreira, P., *New Scientist*, 12 October (2010).

Fock, V., *Voprosy Philosophii*, 1, 168 (1953).

Fowler, R., *Mon. Not. Roy. Ast. Soc.*, 87, 114 (1926).

Friedan, D., http://arxiv.org/abs/hep-th/0204131 (2002).

Friedmann, A., *Zeitschrift für Physik*, 10, 377 (1922).

Gamow, G., *Nature*, 162, 680 (1948).

Garwin, R., *Physics Today*, 27, 9 (1974).

Giacconi, R., et al., *Phys. Rev. Lett.*, 9, 439 (1962).

Gibbs, G., *Sci. Am.*, April, 89 (2002).

Giddings, S., http://arxiv.org/abs/1105.6359v1 (2011a).

———, http://arxiv.org/abs/1108.2015v2 (2011b).

Glanz, J., *Science*, 279, 651 (1998).

Gödel, K., *Rev. Mod. Phys.*, 21, 447 (1949).

Goenner, H., *Liv. Rev. Rel.*, 7 (2004).

Gorelik, G., *Sci. Am.*, August, 72 (1997).

Green, M., and J. Schwarz, *Phys. Lett. B*, 149, 117 (1984).

Greenstein, J., *Ann. Rev. Astron. Astroph.*, 22, 1 (1984).

Gross, D., *Nuc. Phys. B.*, 236, 349 (1984).

Guth, A., *Phys. Rev. D*, 23, 347 (1981).

Guzzo, L., et al., http://arxiv.org/abs/0802.1944 (2008).

Hamber, H., http://arxiv.org/abs/0704.2895v3 (2007).

Hanany, S., *Astroph. Jour. Lett.*, 545, 5 (2000).

Hanbury-Brown, R., *IAU Supp.*, 9, 471B (1959).

Hannam, M., *Class. Quant. Grav.*, 26, 114001 (2009).

Harvey, A., and E. Schucking, *Am. Journ. Phys.*, 68, 723 (1999).

Harwitt, M., interview with P. J. E. Peebles for AIP, http://www.aip.org/history/ohilist/4814.html (1984).

Hawking, S., *Phys. Rev. Lett.*, 17, 444 (1966).

———, *Comm. Math. Phys.*, 25, 152 (1971a).

———, *Phys. Rev. Lett.*, 26, 1344 (1971b).

———, *Nature*, 248, 30 (1974).

———, *Comm. Math. Phys.*, 43, 199 (1975).

———, *Phys. Rev. D*, 13, 13 (1976a).

———, *Phys. Rev. D*, 14, 2460 (1976b).

———, *Nuc. Phys. B*, 144, 349 (1978).

———, *Comm. Math. Phys.*, 87, 395 (1982).

Hawking, S., and G. Ellis, *Astroph. Jour.*, 152, 25 (1968).

Hawking, S., and R. Penrose, *Proc. Roy. Soc. Lon. A*, 314, 529 (1970).

Hegyi, D., ed., 6th Texas Symposium on Relativistic Astrophysics, *Ann. New York Ac. Sci.*, 224 (1973).

Hetherington, N., *Nature*, 316, 16 (1986).

Hewish, A., S. Bell, J. Pilkington, P. Scott, and R. Collins, *Nature*, 217, 709 (1968).

Hoyle, F., *Mon. Not. Roy. Ast. Soc.*, 108, 372 (1948).

Hoyle, F., and G. Burbidge, *Astroph. Jour.*, 144, 534 (1966).

Hoyle, F., and J. Narlikar, *Proc. Roy. Soc. Lon. A*, 273, 1 (1963).

Hoyt, W., Biographical Memoirs, *Nat. Ac. Sci.* 52, 411 (1980).

Hubble, E., *Astr. Jour.*, 64, 321 (1926).

———, *Astr. Jour.*, 69, 103 (1929a).

———, *Proc. Nat. Ac. Sci.*, 15, 168 (1929b).

Hughes, S., http://arxiv.org/abs/hep-ph/0511217 (2005).

Humason, M., *Proc. Nat. Ac. Sci.*, 15, 167 (1929).

Huterer, D., and M. Turner, http://arxiv.org/abs/astro-ph/9808133 (1998).

Ioffe, B., http://arxiv.org/abs/hep-ph/0204295 (2002).

Isham, C., http://arxiv.org/abs/gr-qc/9210011 (1992).

Israel, W., *Phys. Rev.*, 164, 1776 (1967).

Jacobson, T., http://arxiv.org/abs/gr-qc/9908031 (1999).

Jacobson, T., and L. Smolin, *Nuc. Phys. B*, 299, 295 (1988).

Jansky, K., *Proc. IRE*, 21, 1387 (1933).

Janssen, M., University of Minnesota Colloquium at https://sites.google.com/a/umn.edu/micheljanssen/home/talks (2006).

Jennison, R., and M. Das Gupta, *Nature*, 172, 996 (1953).

Kennefick, D., *Physics Today*, September, 43 (2005).

Kerr, R., *Phys. Rev. Lett.*, 11, 237 (1963).

Kragh, H., *Centaurus*, 32, 114 (1987).

Kragh, H., and R. Smith, *Hist. Sci.*, 41, 141 (2003).

Krasnov, K., http://arxiv.org/abs/gr-qc/9710006 (1997).

Landau, L., *Physikalische Zeitschrift der Sowjetunion*, 1, 258 (1932).

———, *Nature*, 364, 333 (1938).

Lemaître, G., *Ann. de la Soc. Sci. de Brux.*, A47, 49 (1927).

———, *Nature*, 127, 706 (1931).

———, *Proc. Nat. Ac. Sci.*, 20, 12 (1934).

———, *Ricerche Astronomiche*, 5, 475 (1958).

Lenard, P., Nobel lecture, http://www.nobelprize.org/nobel_prizes/physics/laureates/1905 (1906).

Le Verrier, U., *Ann. De l'Obs. Imp. Paris*, IV (1858).

Lifshitz, E., and I. Khalatnikov, *Soviet Physics — JETP*, 12, 108 and 558 (1961).

Lightman, A., interview with G. de Vaucouleurs for AIP, http://www.aip.org/history/ohilist/33930.html (1988a).

———, interview with P. J. E. Peebles for AIP, http://www.aip.org/history/ohilist/33957.html (1988b).

Linde, A., *Phys. Lett. B*, 108, 389 (1982).

Lundmark, K., *Mon. Not. Roy. Ast. Soc.*, 84, 747 (1924).

Lynden-Bell, D., *Nature*, 223, 690 (1969).

Lynden-Bell, D., and M. Rees, *Mon. Not. Roy. Ast. Soc.*, 152, 461 (1971).

Maksimov, A., *Red Fleet*, 14 June (1952).

Mathur, S., http://arxiv.org/abs/gr-qc/0502050 (2005).

———, http://arxiv.org/abs/0909.1038v2 (2009).

Milgrom, M., *Astroph. Jour.*, 270, 365 (1983).

Mills, B., and O. Slee, *Aust. Jour. Phys.*, 10, 162 (1956).

Misner, C., *Rev. Mod. Phys.*, 29, 497 (1957).

———, *Astrophys. Space Sci. Lib.*, 367, 9 (2010).

Mooallem, J., *Harper's Magazine*, October, 84 (2007).

Mota, E., P. Crawford, and A. Simões, *Brit. Journ. Hist. Sci.*, 42, 245 (2008).

Neyman, J., and E. Scott, *Astroph. Jour.*, 116, 144 (1952).

———, *Astroph. Jour. Supp.*, 1, 269 (1954).

Norton, J., in *Reflections on Spacetime*, Kluwer Academic Publishing (1992).

———, *Stud. Hist. Phil. Mod. Phys.*, 31, 135 (2000).

Novikov, I., *Soviet Ast.*, 11, 541 (1967).

Nussbaumer, H., and L. Bieri, http://arxiv.org/abs/1107.2281 (2011).

Oppenheimer, J. R., and R. Serber, *Phys. Rev.*, 54, 540 (1938).

Oppenheimer, J. R., and H. Snyder, *Phys. Rev.*, 56, 455 (1939).

Oppenheimer, J. R., and G. Volkoff, *Phys. Rev.*, 55, 375 (1939).

Osterbrock, D., R. Brashear, and J. Gwinn, *Ast. Soc. Pac.*, 10, 1 (1990).

Ostriker, J., and P. Steinhardt, *Nature*, 377, 600 (1995).

Overbye, D., *New York Times*, November 11, 2003.

Peacock, J., http://arxiv.org/abs/0809.4573 (2008).

Peat, D., and P. Buckley, interview with P. Dirac, http://www.fdavidpeat.com/interviews/dirac.htm (1972).

Peebles, P., *Astroph. Jour.*, 142, 1317 (1965).

———, *Astroph. Jour.*, 146, 542 (1966a).

———, *Phys. Rev. Lett.*, 16, 410 (1966b).

———, *Astroph. Jour.*, 147, 859 (1967).

———, *Nature*, 220, 237 (1968).

———, *Astroph. Jour.*, 158, 103 (1969).

———, *IAU Symp.*, 58, 55 (1974).

———, *Astroph. Jour. Lett.*, 263, 1 (1982).

———, *Astroph. Jour.*, 284, 439 (1984).

———, *Nature*, 327, 210 (1987a).

———, *Astroph. Jour. Lett.*, 315, 73 (1987b).

———, http://arxiv.org/abs/astro-ph/0011252v1 (2000).

———, http://arxiv.org/abs/astro-ph/0410284v1 (2004).

Peebles, P., and J. Yu, *Astroph. Jour.*, 162, 815 (1970).

Penrose, R., *Phys. Rev. Lett.*, 14, 57 (1965).

———, *Nature*, 229, 185 (1971).

Penzia, A., and R. Wilson, *Astroph. Jour.*, 142, 419 (1965).

Perlmutter, S., et al., *Astroph. Jour.*, 517, 565 (1999).

Pretorius, F., *Phys. Rev. Lett.*, 95, 121101 (2005).

———, http://arxiv.org/abs/0710.1338 (2007).

Pringle, J., M. Rees, and A. Pacholczyk, *Astron. & Astroph.*, 29, 179 (1973).

Reber, G., *Astroph. Jour.*, 91, 621 (1940).

———, *Astroph. Jour.*, 100, 279 (1944).

Rees, M., *Mon. Not. Roy. Ast. Soc.*, 135, 145 (1967).

———, *IAU Symposium*, 64, 194 (1974).

———, *The Observatory*, 98, 210 (1978).

Rees, M., and D. Sciama, *Nature*, 207, 738 (1965a).

———, *Nature*, 208, 371 (1965b).

———, *Nature*, 211, 468 (1966).

Reiss, A., et al., *Astroph. Jour.*, 16, 1009 (1998).

Robertson, H., *Proc. Nat. Ac. Sci.*, 93, 527 (1949).

Rovelli, C., http://arxiv.org/abs/gr-qc/9603063 (1996).

———, http://arxiv.org/abs/1012.4707v2 (2010).

Rovelli, C., and L. Smolin, *Phys. Rev. D*, 61, 1155 (1988).

———, *Nuc. Phys. B*, 331, 80 (1990).

———, *Phys. Rev. D*, 52, 5743 (1995).

Rubin, V., *Proc. Nat. Ac. Sci.*, 40, 541 (1954).

———, *Astroph. Jour.*, 159, 379 (1970).

———, *Physics Today*, December, 8 (2006).

Ruffini, R., and J. Wheeler, *Physics Today*, January, 30 (1971).

Ryle, M., *The Observatory*, 75, 13 (1955).

Ryle, M., and J. Bailey, *Nature*, 217, 907 (1968).

Ryle, M., and R. Clarke, *Mon. Not. Roy. Ast. Soc.*, 172, 349 (1961).

Ryle, M., F. Smith, and B. Elsmore, *Mon. Not. Roy. Ast. Soc.*, 110, 508 (1950).

Sachs, R., and A. Wolfe, *Astroph. Jour.*, 147, 73 (1967).

Sakharov, A., *Nature*, 331, 671 (1988).

Salpeter, E., *Astroph. Jour.*, 140, 796 (1964).

Schucking, E., *Physics Today*, August, 46 (1989).

———, http://arxiv.org/abs/0903.3768 (2009).

Sciama, D., *Nature*, 224, 1263 (1969).

Sciama, D., G. Field, and M. Rees, *Phys. Rev. Lett.*, 23, 1514 (1969).

Sciama, D., and M. Rees, *Nature*, 211, 1283 (1966).

Shapiro, B., interview with M. Humason for AIP, www.aip.org/history/ohilist/4686.html (1965).

Shields, G., *Pub. Ast. Soc. Pac.*, 111, 661 (1999).

Silk, J., *Astroph. Jour.*, 151, 459 (1968).

Slipher, V., *Lowell Observatory Bulletin*, 58 (1913).

———, *Lowell Observatory Bulletin*, 62 (1914).

———, *Proc. Amer. Phil. Soc.*, 56, 403 (1917).

Smeenk, C., interview with P. J. E. Peebles for AIP, http://www.aip.org/history/ohilist/25507_1.html (2002).

Smolin, L., *Nuc. Phys. B*, 160, 253 (1979).

Smoot, G., et al., *Astroph. Jour. Lett.*, 396, 1 (1992).

Stelle, K., http://arxiv.org/abs/hep-th/0503110v1 (2005).

———, *Nature Physics*, 3, 448 (2007).

———, *Fortschr. Phys.*, 57, 446 (2009).

Stoner, E., *Philosophical Magazine*, 7, 63 (1929).

Straumann, N., http://arxiv.org/abs/gr-qc/0208027 (2002).

Strominger, A., *Nuc. Phys. B*, 192, 119 (2009).

Strominger, A., and C. Vafa, *Phys. Lett. B*, 379, 99 (1996).

Susskind, L., http://arxiv.org/abs/hep-th/9309145v2 (1993).

Susskind, L., and L. Thorlacius, http://arxiv.org/abs/hep-th/9308100v1 (1993).

Susskind, L., L. Thorlacius, and J. Uglum, http://arxiv.org/abs/hep-th/9306069v1 (1993).

't Hooft, G., *Nuc. Phys. B*, 256, 727 (1985).

———, *Nuc. Phys. B*, 335, 138 (1990).

———, http://arxiv.org/abs/gr-qc/9310026v2 (1993).

———, http://arxiv.org/abs/hep-th/0003004v2 (2000).

Thorne, K., *LIGO Report*, P-000024-00-D (2001).

Tolman, R., *Phys. Rev. D*, 55, 364 (1939).

Trimble, V., *Beam Line*, 28, 21 (1998).

Tyson, A., and R. Giffard, *Ann. Rev. Astron. Astroph.*, 16, 521 (1978).

Unzicker, A., http://arxiv.org/abs/0708.3518 (2008).

van den Bergh, S., http://arxiv.org/abs/astro-ph/9904251 (1991).

Vittorio, N., and J. Silk, *Astroph. Jour.*, 297, L1 (1985).

Wang, L., et al., *Astroph. Jour.,* 530, 17 (2000).

Wazak, M., *New Scientist,* November, 27 (2010).

Weart, S., interview with S. Chandrasekhar for AIP, http://www.aip.org/history/ohilist/4551_3.html (1977).

——, interview with T. Gold for AIP, http://www.aip.org/history/ohilist/4627.html (1978).

Weber, J., *Phys. Rev. Lett.,* 22, 1320 (1969).

——, *Phys. Rev. Lett.,* 24, 276 (1970a).

——, *Phys. Rev. Lett.,* 25, 180 (1970b).

——, *Nature,* 240, 28 (1972).

Weber, J., and J. Wheeler, *Rev. Mod. Phys.,* 29, 509 (1957).

Weinberg, S., *Phys. Rev.,* 138, 988 (1965).

——, *Phys. Rev. Lett.,* 59, 2607 (1987).

Weyl, H., *Zeitschrift für Physik,* 24, 230 (1923).

Wheeler, J., *Phys. Rev.,* 97, 511 (1955).

——, *Phys. Rev.,* 2, 604 (1957).

——, *Ann. Rev. Astron. Astroph.,* 4, 393 (1966).

White, S., et al., *Nature,* 330, 451 (1987).

Wick, G., *Physics Today,* February, 1237 (1970).

Williamson, R., *Jour. Roy. Astron. Soc. Can.,* 45, 185 (1951).

Witten, E., *Physics Today,* April, 24 (1996a).

——, *Nature,* 383, 215 (1996b).

——, *Notices of the AMS,* 45, 1124 (1998).

Woodard, R., http://arxiv.org/abs/0907.4238 (2009).

Wright, P., interview with M. Schmidt for AIP, http://www.aip.org/history/ohilist/4861.html (1975).

Zel'dovich, Y., *Soviet Physics — Doklady,* 9, 195 (1964).

——, *Soviet Physics Uspekhi,* 11, 381 (1968).

——, *JETP Letters,* 14, 180 (1971).

——, *Mon. Not. Roy. Ast. Soc.,* 160, 7 (1972).

Zel'dovich, Y., and O. Guseinov, *Astroph, Jour.,* 144, 840 (1965).

Zel'dovich, Y., and A. Starobinsky, *Soviet Physics — JETP,* 34, 1159 (1972).

Index

acceleration: in expanding universe model, 189–90, 216
  gravity and, 9–10, 13–14
accretion disks, 132–33, 135
"Against the Ignorant Criticism of Modern Theories in Physics" (Fock), 72
"Against the Reactionary Einsteinianism in Physics" (Maximow), 71–72
Andromeda Galaxy, 39
  identification of, 40–41
  Rubin and, 182–83
anthropic principle: Carter and, 231
  Dicke and, 230–31
"Appeal to Europeans, An" (Einstein), 17
Arecibo Observatory (Puerto Rico), 99
Ashtekar, Abhay: solves field equations, 201, 203
astronomy: general relativity's effects on, 29
  Newtonian physics in, 7–9
  use of gravity waves in, 160–62, 168
atomic bomb, xvi, 63–64, 87, 101–2
  German attempt to develop, 69
  Soviet Union develops, 72, 121–23, 213–14

Baade, Walter, 179–80
Babson, Roger: and gravity, 105–6, 107–8
Bahcall, John, 169
Bahnson, Agnew: and gravity, 108
Bardeen, James, 133–34

"Beginning of the World From the Point of View of the Quantum Theory, The" (Lemaître), 44–45
Bekenstein, Jacob: and black holes, 146–48, 151, 193, 217, 219
  and field equations, 219
  and modifications to general relativity, 217–18, 221
  and Modified Newtonian Dynamics, 218, 219
  and tensor-vector-scalar theory, 219–20
Belinski, Vladimir, 125
Bell, Jocelyn, xiv, 229
  denied Nobel Prize, 130
  discovers pulsars, 129–30
  "Observation of a Rapidly Pulsating Radio Source," 129
  and quasars, 129
Bergmann, Peter, 120, 197
Beria, Lavrentiy, 72, 213
Bethe, Hans, 63
"Big Bang Cosmology—Enigmas and Nostrums, The" (Dicke & Peebles), 180–81, 182
"Big Bang" theory: Dicke and, 176
  Eddington rejects, 44–45, 90
  Einstein accepts, 46
  and expanding universe model, 181
  Gamow and, 127
  and general relativity, 216

Hawking and, 128, 229
Hoyle rejects, 85, 126–27
Lemaître proposes, 44–45, 59, 85, 88, 90,
    126, 181
observational evidence for, 127, 176
Peebles and, 176, 180–81, 182, 188, 216
Penrose and, 229
Penzias and, 126–27, 128, 176, 178
Rees and, 173
relic radiation in, 176, 178–79, 182, 188–89
Sachs and, 178
Sciama accepts, 126, 127
Silk and, 178, 182, 188
string theory and, 229–30
Wilson and, 126–27, 128, 176, 178
Wolfe and, 178
Zel'dovich and, 178, 181, 182, 188
Big Bang Theory (TV series), 205
"Black Hole Information Paradox" (Hawking),
    195–96, 205, 206
black holes, xiv–xv. *See also* stars, evolution
        and decay of
    and accretion disks, 132–33
    area of, 146, 148, 151, 207
    Bekenstein and, 146–48, 151, 193, 217, 219
    Carter and, 133–34, 146
    at center of Milky Way, 227–28
    Chandra on, 57, 64
    collision of, 163–65, 169–70, 225
    Cygnus X-1, 133
    DeWitt and, 133–34, 150, 164–65
    Eddington rejects, 124, 125, 134, 226
    Einstein rejects, 64–65, 124, 125, 134, 226
    energy production by, 131–32
    entropy of, 146–48, 151, 193, 196, 206
    event horizon of, 147, 148, 205, 225
    and general relativity, 118, 123–24, 131–32,
        134, 205, 226
    Hawking and, 128, 133–34, 135, 145–46,
        148–51, 193–95
    and information paradox, 196, 205, 206–7
    Israel and, 146
    Kerr and, 131
    Khalatnikov and, 123–24, 125
    Lifshitz and, 123–24, 125
    Lynden-Bell and, 132–33, 225, 226
    nature of, 132–35

    Newman and, 134
    Novikov and, 134, 135, 225, 226, 227
    observational evidence for, 132–33, 226–27
    Oppenheimer and Snyder study, 47–48, 49,
        64, 120, 123
    Penrose and, 124–26, 128, 132, 135, 145–46,
        148, 226
    quantum physics and, 148–49
    quasars as, 228
    radiation by, 145, 149–51, 193–95, 196, 205,
        206–7
    as radio sources, 131–32
    Rees and, 135, 173, 225, 226
    satellite-based study of, 224–25
    Schwarzschild discovers, 49–50, 64, 75, 131,
        132
    Sciama and, 150
    and search for gravity waves, 163–64,
        169–70, 225
    Smarr and, 165, 166
    Thorne and, 133–34, 135, 226
    Wheeler and, 100, 103, 116, 118, 123, 132, 134,
        150, 175, 226
    Zel'dovich and, 131–32, 133, 135, 148, 225,
        226, 227
Blumenthal, George: and cosmological
        constant, 187
Bohr, Niels, 73, 122
    and quantum physics, 101
Bohr, Niels & John Archibald Wheeler: "The
        Mechanism of Nuclear Fission," 62–63
Boltzmann, Ludwig, 3
Bondi, Hermann, 86, 87–88, 114, 120
    and general relativity, 98, 119, 154
    and gravity waves, 154
    performs thought experiments, 154
    and steady-state theory, 89–91, 96, 97–98,
        119
Born, Max, 58
    flees Germany, 69
    on steady-state theory, 91
Brans, Carl: and gravity, 212–13
Bronstein, Matvei, 143
Burbidge, Geoffrey & Margaret, 98
    and stellar energy sources, 113
Burnell, Jocelyn. *See* Bell, Jocelyn
Butterfield, Herbert, 91

Calabi-Yau geometry: and string theory, 199
Cambridge, University of: cosmology at,
    86–93, 94, 96–98, 119
Candelas, Philip, 144
  on Hawking, 149–50
  and string theory, 199, 204
Carter, Brandon: and anthropic principle, 231
  and black holes, 133–34, 146
CDM model of universe, 184–85. *See also*
    dark matter
  cosmological constant in, 185, 186–87, 189,
    191
  and galaxy formation, 184, 185
  inflationary model and, 185
  large-scale structure theory and, 184, 189
  Peebles and, 184–85, 188, 191
Center for Relativity (Austin): Kerr at, 120–21
  Penrose at, 120–21
  Schild forms, 114–15, 120
Cepheids, 40
CERN (Geneva): and unified theory, 142, 144
Chandrasekhar, Subrahmanyan: on black
    holes, 57, 64
  Eddington opposes, 57, 64, 211
  and Gödel, 79
  and quantum physics, 54–58, 59
  and radio astronomy, 94
  and stellar evolution and decay, 55–57, 64,
    115, 129, 139, 211
cold dark matter model. *See* CDM model of
  universe
Coleman, Sidney, 201
concordance model of universe, 190–92
  large-scale structure theory and, 191
Copernicus, Nicolaus, 29
Cosmic Background Explorer (satellite):
    measures relic radiation, 188
  Smoot and, 188
"Cosmic Static" (Reber), 94, 95
"cosmic web," 180, 224
cosmological constant: Blumenthal and, 187
  in CDM model, 185, 186–87, 189, 191
  dark energy and, 192, 216
  Davis and, 186
  Dekel and, 187
  discarded, 43, 185–86
  Eddington and, 185–86

Efstathiou and, 210
Einstein introduces, 29–30, 32–33, 37–38,
    174–75, 185
energy and, 186
Frenk and, 186
Gott and, 187
High-Z Supernova Search project and,
    189–90
Lemaître and, 186
in multiverse model, 230–31
observational evidence for, 187, 190–91
Ostriker and, 187
Peebles and, 175, 187, 191
Primack and, 187
quantum gravity and, 192
rehabilitation of, 189–92
Schmidt and, 190
Spergel and, 187
Steinhardt and, 187
in string theory, 231
Supernova Cosmology Project and, 189–90
supernovae and, 189
Turner and, 187
White and, 186
Zel'dovich and, 186, 191, 192
"Cosmological Constant and Cold Dark
    Matter, The" (Efstathiou), 186–87
cosmology: concordance model, 190–92
  cyclical model, 230
  expanding universe model, 30–31, 33–34,
    36–39, 41–46, 48, 71, 88–89, 96, 177
  Gott on, 174
  Hoyle popularizes, 85–86, 91–92
  inflationary universe model, 181–82
  large-scale structure theory in, 174–75,
    179–80
  multiverse model, 230–31
  nature of, 173–74, 190
  observational evidence in, 220–22, 231
  Peebles and, 176–77
  popular interest in, 228
  radio astronomy and, 97, 234–35
  Rees on, 173–74
  Sakharov and, 213–14
  Spergel on, 174
  static universe model, 29–30, 32–35, 36, 45,
    85–86, 90, 92

steady-state theory in, 85–86, 89–92, 96
  Turner on, 174–75
  at University of Cambridge, 86–93, 94,
    96–98, 119
Cosmos (TV series), x
Cottingham, Edward, 24
Course of Theoretical Physics (Landau), 72
"creation field": Hoyle and, 90–91, 110
Critical Dialogues in Cosmology (Princeton,
    1996), 173–75, 187
Crommelin, Andrew, 24–25
cyclical model of universe, 230
Cygnus A: as radio source, 112
Cygnus X-1: as black hole, 133

dark energy, xiv–xv, 174, 220–21, 224
  and cosmological constant, 192, 216
dark matter, xiv, xv, 174. *See also* CDM model
    of universe
  distribution of, 184–85, 218
  Faber and, 183–84
  Gallagher and, 183–84
  general relativity and, 217
  in large-scale structure theory, 209–10, 224
  Modified Newtonian Dynamics and, 218–19
  observational evidence for, 233
  Peebles and, 184
  standard model and, 174
Davidson, Charles, 24–25
Davis, Marc, 180
  and cosmological constant, 186
  and galaxy formation, 184
de Sitter, Willem, 22
  on age of universe, 89
  and expanding universe model, 31, 33–34,
    36–37, 38–39, 41, 42–43
  and static universe model, 30, 37
de Sitter effect, 37, 38, 41–43, 88
de Vaucouleurs, Gérard: and galaxy
    superclusters, 179–80
Dead of Night (film), 90
Dekel, Avishai: and cosmological constant, 187
Deser, Stanley, 144, 201
"Detection of Gravity Waves Challenged"
    (Garwin), 160
"Detective Police, The" (Dickens), 7–8

DeWitt, Bryce: background of, 105
  and black holes, 133–34, 150, 164–65
  and canonical approach to spacetime, 197,
    201
  and gravitons, 143–44, 150, 165, 197
  and gravity, 107–108
  and Institute of Field Physics, 108, 110
  and quantum electrodynamics, 143, 197
  and quantum gravity, 137, 143–44, 197–98,
    201, 203
  on string theory, 203
  Wheeler and, 197
DeWitt-Morette, Cécile, 105, 108, 133, 201
"Dialectical and Historical Materialism"
    (Stalin), 70
dialectical materialism: general relativity and,
    70–71
  quantum physics and, 70–71
Dicke, Robert, 180, 188
  and anthropic principle, 230–31
  and "Big Bang" theory, 176
  and general relativity, 111–12, 175–76, 212
Dicke, Robert & Philip Peebles: "The Big
    Bang Cosmology—Enigmas and
    Nostrums," 180–81, 182
Dickens, Charles: "The Detective Police,"
    7–8
Dirac, Paul, xi, 18, 143, 197
  background and personality of, 138, 142
  on field equations, 212
  and gravity, 212–13
  as Lucasian Professor of Mathematics, 138,
    142, 150, 193
  and mathematics, 138, 140–41, 142
  and modifications to general relativity, 213,
    217–18, 221
  and quantum electrodynamics, 140–41
  and quantum physics, 119, 138–43
  Sciama and, 119–20
  and standard model, 142
  and unified theory, 138–39
  wins Nobel Prize, 140
Dirac equation, 138–40
Doppler effect, 38
Drever, Ronald, 167
Duff, Michael: and gravity, 144–45
  on string theory, 203–4

Dyson, Frank, 23, 232
Dyson, Freeman: on general relativity, 81
    and Weber, 170–71

Eddington, Sir Arthur, 87–88
    academic career, 17–18
    and clumping of matter, 45
    and cosmological constant, 185–86
    and expanding universe model, 36, 37, 39,
        41, 44–45
    The Fundamental Theory, 86–87
    The Internal Constitution of the Stars, 50–51,
        54, 55, 61, 87
    leads Príncipe expedition (1919), ix, 22–25,
        26, 30, 34, 50, 111, 211, 232–33, 234–35
    and Lemaître, 36, 43
    and nuclear fusion, 62
    opposes Chandra, 57, 64, 211
    personality of, 17
    political opinions, 18–19, 22–23, 35, 42
    promotes general relativity, ix–xi, 18, 22–23,
        26, 30, 50, 64, 86
    rejects "Big Bang" theory, 44–45, 90
    rejects black holes, 124, 125, 134, 226
    rejects gravity waves, 154
    on Schwarzschild, 48
    Stars and Atoms, 87
    and stellar evolution and decay, 50–52,
        54–55, 56–57, 64, 65, 114
    and unified theory, 86–87, 137, 212
"Effect of Wind and Air-Density on the Path of
    a Projectile, The" (Schwarzschild), 48
Efstathiou, George: and cosmological constant,
    210
    "The Cosmological Constant and Cold
        Dark Matter," 186–87
    and galaxy formation, 184
    and large-scale structure theory, 209–10
    and standard model, 221–22
Ehrenfest, Paul, 31
Eidgenössische Technische Hochschule Zürich
    (ETH): Einstein as professor at, 14
Einstein, Albert: accepts "Big Bang" theory, 46
    "An Appeal to Europeans," 17
    background of, 2–3, 26–27
    as celebrity, 67–68, 234

death of, 83
and distribution of matter, 78
divorces Marić, 16
emigrates to United States, 68–70, 73, 75
and expanding universe model, 43, 44,
    45–46, 90
as fellow of Prussian Academy of Sciences,
    16, 21
and gravity waves, 153–54
Grossmann assists, 14–16, 20
heads Kaiser Wilhelm Institute of Physics,
    16
at Institute for Advanced Study, 66–67, 73,
    119–20
introduces cosmological constant, 29–30,
    32–33, 37–38, 174–75, 185
later life, 66–67
marriage to Lowenthal, 16, 66
marriage to Marić, 3, 11, 16
and mathematics, 12, 14, 20, 22, 73, 76, 199
and McCarthyism, 83
"A New Determination of Molecular
    Dimensions" (1905), 3
"On the Relativity Principle and the
    Conclusions Drawn From It" (1907),
    x, 2, 4, 5, 10, 67
Oppenheimer and, 80, 81–82, 83–84
and orbital decay, 160–61
performs thought experiments, 3–4, 5–6,
    9–10, 22, 71
personality of, 2
and photoelectric effect, 52–53
physical intuition, 16, 20, 29, 31, 46, 73,
    79–80
political opinions, 17, 35, 82
popular acclaim for, 26–27
promotes static universe model, 31, 33–35,
    36, 37, 39, 90, 92
rejects black holes, 64–65, 124, 125, 134, 226
rejects expanding universe model, 31, 33–35,
    36, 39
relationship with Hilbert, 19–20, 22
relationship with Lemaître, 39–40, 45–46,
    64
Schwarzschild and, 48–50
search for unified theory, 73–75, 86, 110, 137,
    211–12

and spacetime, xi, 215
on steady-state theory, 91
and stellar evolution and decay, 64–65
studies non-Euclidian geometry, 14–15,
   20, 22
as Swiss patent inspector, 1, 3–4, 9, 11, 13, 27
as university professor, 11, 13–14
and World War I, 16–17, 19
Einstein, Eduard, 13–14
Einstein, Hans Albert, 11
Einstein, Maja, 66
"Electrical Disturbances Apparently of
   Extraterrestrial Origin" (Jansky), 93
electromagnetism: gravity and, 74, 143
   Hertz and, 153
   Maxwell and, 5–6, 52, 110, 141, 153
   in unified theory, 73, 140–42, 143
electroweak force: in unified theory, 142, 143
Ellis, George, 128
   and general relativity, 231
   on multiverse model, 231–32
energy: and cosmological constant, 186
   distribution of, 28–29, 32–33
   produced by black holes, 131–32
   relationship to mass, x, 2, 50–51
entropy: of black holes, 146–48, 151
Erhard, Werner: sponsors physics lectures, 195
European Space Agency, 223, 225–26
Event Horizon Telescope, 227
exclusion principle: Pauli and, 80, 86–87, 139
expanding model of universe, 30–31
   acceleration in, 189–90, 216
   "Big Bang" theory and, 181
   de Sitter and, 30–31, 33–34, 36–37, 38–39,
      42–43
   Eddington and, 36, 37, 39, 44–45
   Einstein and, 43, 44, 45–46, 90
   Einstein rejects, 31, 33–35, 36
   Friedmann and, 33–35, 36, 37, 39, 42–45, 46,
      48, 71, 88–90, 91–92, 96, 177
   gravity and, 177–78
   Hoyle and, 88, 89
   Lemaître and, 37–38, 39, 42–46, 48, 88–90,
      91–92, 96, 177
   Lifshitz and, 72
   and redshift effect, 36–37, 38–39, 41–42
   Weyl and, 37, 39

Faber, Sandra: and dark matter, 183–84
Faraday, Michael, 1
Fermi, Enrico, 141
Feynman, Richard, 63, 143, 195
   and general relativity, 109–10, 111
   and gravity waves, 154
   and quantum electrodynamics, 140
   and quantum physics, 102, 109
   on string theory, 200
   Wheeler and, 102
   wins Nobel Prize, 140
Field, George: and gravity waves, 158
field equations, 21, 28–29, 30, 111, 195
   Ashtekar solves, 201, 203
   Bekenstein and, 219
   Dirac on, 212
   Friedmann and, 32–33, 64, 67, 78, 80
   Gödel solves, 78–80
   gravity waves and, 164, 165
   Hoyle and, 90
   Kerr solves, 120–21, 125, 134, 148
   Lemaître and, 67, 78, 80
   Lifshitz and, 177–78
   Pretorius solves, 169–70
   Sakharov and, 214
   Schwarzschild solves, 48–50, 51–52, 58,
      61–62, 64, 67, 80, 116, 120–21, 125, 134
   in unified theory, 74
"final state." *See* stars, evolution and decay of
Flerov, Georgii, 122
Fock, Vladimir: "Against the Ignorant
   Criticism of Modern Theories in
   Physics," 72
Fowler, Ralph, 18
   and stellar energy sources, 113–14
   and stellar evolution and decay, 55–56, 139
Fowler, William, 98
Frenk, Carlos: and cosmological constant, 186
   and galaxy formation, 184
Friedan, Daniel: on string theory, 200
Friedmann, Alexander, 28, 111
   background and personality of, 32
   death of, 35
   and expanding universe model, 33–35, 36,
      37, 39, 42–45, 46, 48, 71, 88–90, 91–92,
      96, 177
   and field equations, 32–34, 64, 78, 80

Friedmann, Alexander (*cont.*)
"On the Curvature of Space," 33–34
in World War I, 31–32, 35, 48
"frozen stars." *See* black holes
Fundamental Theory, The (Eddington), 86–87

galaxies: and accretion disks, 132–33
de Vaucouleurs and superclusters of, 179–80
distribution of, 185, 192, 220, 224
formation and structure of, 177–78, 182–84,
185, 216, 218, 220, 233
Gamow and, 177, 178
in general relativity, 177, 220
gravity and, 177–78, 183, 218
and large-scale structure theory, 177
nebulae identified as, 40–41
Zwicky and, 183
Galilei, Galileo, 1
Gallagher, Jay: and dark matter, 183–84
Galle, Gottfried, 7
Gamow, George: and "Big Bang" theory, 127
and galaxies, 177, 178
Garwin, Richard: "Detection of Gravity Waves
Challenged," 160
Gauss, Carl Friedrich: and non-Euclidian
geometry, 14–15, 21
Gell-Mann, Murray: and string theory, 198–99
general theory of relativity. *See* relativity,
general
geometry, non-Euclidian: Einstein studies,
14–15, 20, 22
Gauss and, 14–15, 21
Riemann and, 15, 21, 74
geometry, quantum, 201
German Charles-Ferdinand University
(Prague): Einstein as professor at, 11
Germany: Jewish physicists flee, 63, 68–69
Giaccone, Riccardo, 133
Glashow, Sheldon: and electroweak force, 142
and standard model, 144
on string theory, 200
Gödel, Kurt: background of, 76
Chandra and, 79
death of, 80
emigrates to United States, 77
and general relativity, 75–76, 78–80

and incompleteness theorem, 76–77
at Institute for Advanced Study, 77
and mathematics, 76–78
solves field equations, 78–80
and time travel, 78–79
Gold, Thomas, 86, 87–88, 95, 114
on general relativity, 117
and radio astronomy, 98–99
and steady-state theory, 89–91, 96, 98, 110
Gott, J. Richard: and cosmological constant,
187
on cosmology, 174
Goudsmit, Samuel: and general relativity, 81
Gravitation and Cosmology (Weinberg), 226,
228
Gravitation (Wheeler, Misner & Thorne), 226
"Gravitational Collapse and Spacetime
Singularities" (Penrose), 125
gravitons: DeWitt and, 143–44, 150, 165, 197
in quantum gravity, 198, 204
in string theory, 198
gravity: and acceleration, 9–10, 13–14
alternative theories of, 209–10
antigravity research, 105–7
Babson and, 105–6, 107–8
Bahnson and, 108
Brans and, 212–13
deflects light, 12–13, 18, 21, 23–24, 25–26, 50,
75, 232
DeWitt and, 107–8, 137, 143, 197
Dirac and, 212–13
Duff and, 144–45
effect on spacetime, 48–49, 59, 61–62, 227,
233
effect on stars, 51, 55–57, 131–32
and electromagnetism, 74, 143
and expanding universe model, 177–78
and galaxies, 177–78, 183, 218
in general relativity, 6, 9–10, 15–16, 20–22,
29–30, 143–45, 211, 220
Hawking and, 150, 194
Hilbert on, 21–22
Jordan and, 212
Milgrom and modification of, 218
in Newtonian physics, 6–9, 49, 211, 212,
217, 218

quantum theory of. *See* quantum gravity
Sakharov and, 214
supergravity, 194–95, 215
tensor-vector-scalar theory, 219–20
in unified theory, 140–41
wave theory of. *See* gravity waves
as weak force, 218
Gravity Research Foundation: sponsors essay
    competitions, 105–7
gravity waves: black holes and search for,
    163–64, 169–70, 225
Bondi and, 154
detection of, 152–53, 155–57, 158–59, 161–63,
    166–68, 169–72, 225
Eddington rejects, 154
Einstein and, 153–54
Feynman and, 154
Field and, 158
and field equations, 164, 165
general relativity and, 168
Hawking and, 157
laser interferometry and search for, 161–62,
    166–69, 171–72
Milky Way as source of, 157–58, 159
neutron stars and, 160–61, 162–63, 225
Ostriker on, 168
Penrose and, 157
Pretorius and, 165–66, 169–70
Rees and, 158
Sciama and, 158
supernovae and, 163
Thorne and, 167, 169, 171
Tyson on, 168
use in astronomy, 160–62, 168
Weber bars and search for, 155–57, 158–59,
    161–63, 171
Weber searches for, 152–53, 154–60, 163,
    170–71, 172
Green, Michael: as Lucasian Professor of
    Mathematics, 204
and string theory, 199, 204
Grossmann, August, 3
Grossmann, Marcel: assists Einstein, 14–16,
    20
Guth, Alan: and inflationary universe model,
    180

Hawking, Stephen, 142, 215
background of, 128
and "Big Bang" theory, 128, 229
"Black Hole Information Paradox," 195–96,
    205, 206
and black holes, 128, 133–34, 135, 145–46,
    148–51, 193–95
Candelas on, 149–50
and Einstein centenary, 180
on "end of theoretical physics," 194–95
and general relativity, 231
and gravity, 150, 194
and gravity waves, 157
health issues, 128, 146
as Lucasian Professor of Mathematics
    (Cambridge), 150, 193–94, 195, 204
and quantum gravity, 193–95
Sciama and, 127–28, 145
and standard model, 194
on string theory, 203
and supersymmetry, 194–95
Hawking radiation. *See* black holes:
    radiation by
Heisenberg, Werner, 143
and quantum physics, 53–54, 55, 101, 119
Stark attacks, 69
and steady-state theory, 91
uncertainty principle, 53–54, 55
Hertz, Heinrich: and electromagnetism, 153
Hewish, Antony, 129
wins Nobel Prize, 130
High-Z Supernova Search project: and
    cosmological constant, 189–90
Hilbert, David, 37, 58, 66
on gravity, 21–22
and incompleteness theorem, 77
and mathematics, 76–77
relationship with Einstein, 19–20, 22
Horowitz, Gary: and string theory, 199
Hoyle, Fred, 138
background of, 87
and "creation field," 90–91, 110
criticized by colleagues, 86, 91–92
and expanding universe model, 88, 89
and field equations, 90
and Nobel Prize, 98

Hoyle, Fred (*cont.*)
　popularizes cosmology, 85–86, 91–92
　promotes steady-state theory, 85–86, 89–92,
　　96, 97, 98, 113, 119–20, 126, 230
　and quantum physics, 87
　and radar research, 87–88
　rejects "Big Bang" theory, 85, 126–27
　Ryle and, 95
　and stellar energy sources, 87, 113–14
Hubble, Edwin, 59
　background and personality of, 40–41
　Hubble constant, 88
　identifies nebulae as galaxies, 40–41
　measures redshift effect, 41–43, 78, 88–89,
　　111, 189
　"A Relation Between Distance and Radial
　　Velocity Among Extragalactic
　　Nebulae," 42
　in World War I, 40
Hubble Space Telescope, 189, 223
Huchra, John, 180
Hulse, Russell: and neutron stars, 160–61, 162,
　163
Humason, Milton: measures redshift effect,
　41–42, 45, 52, 78, 88, 189
hydrogen bomb, 82, 102, 165

incompleteness theorem: Gödel and, 76–77
　Hilbert and, 77
　Russell and, 77
　Wittgenstein and, 77
inertial frames of reference: in special
　relativity, 4–6, 9, 10, 14, 15
Infeld, Leopold, 114
inflation model of the universe, 181–82, 187
　and CDM model, 185
　effects on spacetime, 182
　Guth and, 181
　large-scale structure theory and, 182
　in unified theory, 180
Institute for Advanced Study (Princeton):
　Einstein at, 66–67, 73, 119–20
　Gödel at, 77
　Oppenheimer directs, 66–67, 80, 83
Institute of Astronomy (Cambridge), 209–10,
　221

Institute of Field Physics: DeWitt and, 108, 110
　inaugural meeting (Chapel Hill, 1957),
　　109–11, 114, 115, 154, 221
Internal Constitution of the Stars, The
　(Eddington), 50–51, 54, 55, 61, 87
International Astronomical Union, 216, 218, 232
Israel, Werner: and black holes, 146
　and Einstein centenary, 180

Jacobson, Theodore: and quantum gravity, 201
Jansky, Karl: "Electrical Disturbances
　　Apparently of Extraterrestrial Origin,"
　　93
Jewish physicists: flee Germany, 63, 68–69
"Jewish physics": Nazis oppose, 68–70, 75
Jodrell Bank Observatory (Manchester), 93, 97
Jordan, Pascual: and gravity, 212

Kaiser Wilhelm Institute of Physics (Berlin):
　Einstein heads, 16
Kaluza, Theodor: and five-dimensional
　universe, 73–74
Kennedy assassination (1963), 115
Kerr, Roy: and black holes, 131
　at Center for Relativity, 120–21
　solves field equations, 120–21, 125, 134, 148
Khalatnikov, Isaak: and black holes, 123–24, 125
Klein, Oskar: and five-dimensional universe,
　73–74
Krasnow, Kirill: and quantum gravity, 206
Kurchatov, Yakov, 213

lambda. *See* cosmological constant
Landau, Lev Davidovich, 213
　Course of Theoretical Physics, 72
　and nuclear physics, 122–23
　political opinions, 60
　and quantum physics, 59, 122
　and stellar energy sources, 59–61
　and stellar evolution and decay, 104, 130, 139
Large Hadron Collider (LHC): popular fear
　of, 228–29
large-scale structure theory of universe: and
　CDM model, 184, 188
　and concordance model, 191

in cosmology, 174–75, 179–80
dark matter in, 209–10, 224
Efstathiou and, 209–10
galaxies and, 177
and inflation model, 182
Peebles and, 175, 177, 179–80, 187–88, 189, 192, 216
Silk and, 187
Zel'dovich and, 187
Laser Interferometer Gravitational Wave Observatory (LIGO)
funding and construction of, 167–69, 170, 224, 225
Laser Interferometer Space Antenna (LISA), 224, 225–26
laser interferometry: GEO600, 167–68, 224
satellite-based, 224
and search for gravity waves, 161–62, 166–69, 171–72
TAMA, 167
Weber and, 161
Le Verrier, Urbain: and discovery of Neptune, 7–8
and Mercury, 8–9, 20–21, 217
Lemaître, Georges, 28, 111
"The Beginning of the World From the Point of View of the Quantum Theory," 44–45
and cosmological constant, 186
Eddington and, 36, 43
and expanding universe model, 37–38, 39, 42–46, 48, 88–90, 91–92, 96, 177
and field equations, 67, 78, 80
as Jesuit, 35
promotes general relativity, 36, 71
proposes "Big Bang" theory, 44–45, 59, 85, 88, 90, 126, 181
relationship with Einstein, 39–40, 45–46, 64
in World War I, 35
Lenard, Philipp: discovers photoelectric effect, 52
rejects general relativity, 68
wins Nobel Prize, 68
Lifshitz, Evgeny: and black holes, 123–24, 125
and expanding universe model, 72
and field equations, 177–78

light: gravity deflects, 12–13, 18, 21, 23–24, 25–26, 50, 75, 233
in Newtonian physics, 52
particle theory of, x, 2, 53
in quantum physics, 52–53
and redshift effect, 36–37, 38–39, 41–43, 45, 52, 61, 78
relationship to mass, 103
speed of, 5–6, 9, 10, 49, 56, 154
loop quantum gravity, 201–2, 203–4, 206
Lovell, Bernard: and radio astronomy, 93, 94
and radio sources, 112
Lowell Observatory (Flagstaff), 38
Lowenthal, Elsa: marriage to Einstein, 16, 66
Lucasian Professors of Mathematics (Cambridge): Dirac as, 138, 142, 150, 193
Green as, 204
Hawking as, 150, 193–94, 195, 204
Newton as, 138, 193
Lundmark, Knud: measures distance of nebulae, 39, 41, 42
Lynden-Bell, Donald: and black holes, 132–33, 225, 226

Manhattan Project, 122
Oppenheimer and, 63–64, 81–82
Wheeler and, 101
Marić, Mileva: Einstein divorces, 16
marriage to Einstein, 3, 11, 16
mass: relationship to energy, x, 2, 50–51
relationship to light, 103
mathematics: Dirac and, 138, 140–41, 142
Einstein and, 12, 14, 20, 22, 73, 76, 199
Gödel and, 76–78
Hilbert and, 76–77
renormalization in, 140–41, 142, 143, 145, 194, 198
matter: distribution of, 28–29, 32–33
Eddington and clumping of, 45
Einstein and distribution of, 78
Maximow, Alexander: "Against the Reactionary Einsteinianism in Physics," 71–72
Maxwell, James Clark, 1
and electromagnetism, 5–6, 52, 110, 141, 153

McCarthyism: Einstein and, 83
    Oppenheimer and, 82–83, 102
Meeting on General Relativity and Cosmology,
        Third (London, 1965), 124
Mercury: Le Verrier and, 8–9, 20–21, 217
    orbit of, 10, 26, 49–50
Milgrom, Mordehai: and gravity, 218
    and Modified Newtonian Dynamics, 218
Milky Way galaxy, 29, 40
    black hole at center of, 227–28
    radio astronomy and, 93–94, 95
    as source of gravity waves, 157–58, 159
Mills, Bernard: and radio sources, 96–97
Milne, E. A.: on steady-state theory, 91
Minkowski, Hermann, 19
Misner, Charles, 103, 125, 197, 226
Modified Newtonian Dynamics (MOND):
        Bekenstein and, 218, 219
    and dark matter, 218–19
    Milgrom and, 218
    opposition to, 218–19, 221
    Peebles and, 219
Mount Wilson Observatory (Pasadena), 41, 43
multiverse model of universe: cosmological
        constant in, 230–31
    Ellis on, 231–32
    string theory in, 231

National Center for Supercomputing
        Applications, 166
Nature of the Universe, The (radio series),
        85–86, 91–92
Nazis: oppose "Jewish physics," 68–70, 75
nebulae: Hubble identifies as galaxies, 40–41
    Lundmark measures distance of, 39, 41, 42
Neptune: Le Verrier and discovery of, 7–8
Nernst, Walther, 16
neutron stars, 61, 104, 129–30, 139, 226. *See also*
        pulsars
    and general relativity, 160–61
    and gravity waves, 160–61, 162–63, 225
    Hulse and, 160–61, 162, 163
    Taylor and, 160–61, 162, 163
"New Determination of Molecular
        Dimensions, A" (Einstein), 3
Newman, Ezra: and black holes, 134

Newton, Sir Isaac, 1, 26
    as Lucasian Professor of Mathematics, 138,
        193
Newtonian physics: in astronomy, 7–9
    and general relativity, 15, 21, 90–91
    gravity in, 6–9, 49, 211, 212, 217, 218
    inconsistencies and limitations of, 5–6
    invention of, ix
    light in, 52
    prediction in, 53–54
Nobel Prize: Bell denied, 130
    Dirac wins, 140
    Feynman wins, 140
    Hewish wins, 130
    Hoyle and, 98
    Lenard wins, 68
    Penzias & Wilson win, 176
    Perlmutter wins, 190
    Riess wins, 190
    Ryle wins, 99, 130
    Schmidt wins, 190
    Smoot wins, 188
    Townes wins, 161
Novikov, Igor, 132–33, 188, 192
    and black holes, 134, 135, 225, 226, 227
nuclear energy, 59–60
nuclear physics: develops in World War II,
        63–64
    Eddington and fusion, 62
    Landau and, 122–23
    Oppenheimer and fission, 62–64
    Wheeler and, 100, 104, 122
    Zel'dovich and, 122

"Observation of a Rapidly Pulsating Radio
        Source" (Bell), 129
"On the Curvature of Space" (Friedmann),
        33–34
"On the Relativity Principle and the
        Conclusions Drawn From It"
        (Einstein), x, 2, 4, 5, 10, 67
Oppenheimer, J. Robert, 91, 111
    background of, 58
    and black holes, 47–48, 49, 64, 120, 123
    directs Institute for Advanced Study, 66–67,
        80, 83

and Einstein, 80, 81–82, 83–84
and European scientific refugees, 63
and general relativity, 59, 60–62, 80–81
and Manhattan Project, 63–64, 81–82
and McCarthyism, 82–83, 102
and nuclear fission, 62–64
Pauli on, 59
personality of, 58–59
political opinions, 102
and quantum physics, 47, 58–59
and quasars, 115–16
and stellar energy sources, 60–61
and stellar evolution and decay, 61–62, 64,
    75, 81, 100, 101, 104, 115, 116, 130
Ostriker, Jeremiah: and cosmological constant,
    187
and galaxy formation, 183–84
on gravity waves, 168
Oxford Symposium on Quantum Gravity
    (1974), 144–45, 149–50, 197–98

particles: antiparticles, 139–40, 148–49
    gravitons, 143–44
    high-energy, xi
    photons, 53, 139
    and quantum electrodynamics, 140–41
    quarks, 141
    types of, 139–40, 141, 194, 198
Pauli, Wolfgang, xi, 137, 143, 144
    and exclusion principle, 80, 86–87, 139
    on Oppenheimer, 59
Pawsey, Joseph: and radio astronomy, 93, 94
Peebles, Philip James: background and
        personality of, 175–76, 191
    and "Big Bang" theory, 176, 180–81, 182,
        188, 216
    and CDM model, 184–85, 188, 191
    and cosmological constant, 175, 187, 191
    and cosmology, 176–77
    and dark matter, 184
    and evolution of universe, 178–79, 182
    and galaxy formation, 177–78, 183–84, 216,
        220
    on general relativity, 216–17
    and large-scale structure theory, 175, 177,
        179–80, 187–88, 189, 192, 216

and Modified Newtonian Dynamics, 219
Physical Cosmology, 175
Penrose, Roger, 136, 138
    and "Big Bang" theory, 229
    and black holes, 124–26, 128, 132, 135,
        145–46, 148, 226
    at Center for Relativity, 120–21
    creates spin networks, 202
    and general relativity, 118–21, 231
    "Gravitational Collapse and Spacetime
        Singularities," 125
    and gravity waves, 157
Penrose diagrams, 118, 119, 121, 124, 125
Penrose superradiance, 148
Penzias, Arno: and "Big Bang" theory, 126–27,
        128, 176, 178
    wins Nobel Prize, 176
Perlmutter, Saul: wins Nobel Prize, 190
Perrine, Charles, 18
photoelectric effect: Einstein and, 52–53
    Lenard discovers, 52
Physical Cosmology (Peebles), 175
physics, Newtonian. *See* Newtonian physics
physics, quantum. *See* quantum physics
Planck, Max, 16
planets: orbits of, 7–8
plutonium: Seaborg discovers, 63
predictability: general relativity and, 111
Pretorius, Frans: and gravity waves, 165–66,
        169–70
    solves field equations, 169–70
Primack, Joel: and cosmological constant, 187
"primordial egg." *See* "Big Bang" theory
Príncipe expedition (1919): Eddington leads,
        ix, 22–25, 26, 30, 34, 50, 111, 211, 232–33,
        234–35
Principia Mathematica (Whitehead & Russell),
        18
Prussian Academy of Sciences: Einstein as
        fellow of, 16, 21
pulsars, 160. *See also* neutron stars
    Bell discovers, 129–30

quantum electrodynamics (QED), 198
    DeWitt and, 143, 197
    Dirac and, 140–41

quantum electrodynamics (QED) (*cont.*)
  particles and, 140–41
  Schwinger and, 140, 142
  and unified theory, 140–42, 145
quantum geometry, 201
quantum gravity, 145, 150–51, 165, 216, 229
  and cosmological constant, 192
  DeWitt and, 137, 143–44, 197–98, 201, 203
  gravitons in, 198, 204
  Hawking and, 193–95
  Jacobson and, 201
  Krasnow and, 206
  loop quantum gravity, 201–2, 203–4, 206, 230
  Rovelli and, 202, 206
  Smolin and, 201–2, 204, 206
  spin networks and, 202, 203
  string theory and, 199, 202, 206
  Woit and, 204
quantum physics, xi, xv
  basic principles of, 53–54
  and black holes, 148–49
  Bohr and, 101
  Chandra and, 54–58, 59
  and dialectical materialism, 70–71
  Dirac and, 119, 138–43
  exclusion principle in, 55, 80
  Feynman and, 102, 109
  and general relativity, 44, 47–48, 78, 81, 84, 103, 137, 145, 150–51, 194, 197–98, 211, 216–17
  Heisenberg and, 53–54, 55, 101, 119
  Hoyle and, 87
  information and predictability in, 196
  Landau and, 59, 122
  light in, 52–53
  observational evidence for, 137–38
  Oppenheimer and, 47, 58–59
  Schrödinger and, 53, 55, 74, 101, 119, 197
  in Soviet Union, 72–73
  and special relativity, 138–40
  and stars, 50–52, 54–56
  uncertainty principle in, 53–54, 55
  Wheeler and, 101, 207
quarks, 141
quasars, 226. *See also* radio sources
  Bell and, 129

  as black holes, 228
  and general relativity, 116–17, 126, 134
  Oppenheimer and, 115–16
  Rees and, 132–33
  Sciama and, 126, 132–33
  at Texas Symposium (1963), 115–17

radar: developed in World War II, 87–88, 92–93
  Hoyle researches, 87–88, 92
  Ryle researches, 92, 94
radio astronomy: Chandra and, 94, 115
  and cosmology, 97, 234–35
  development of, 93–94, 96
  and general relativity, 234–35
  Gold and, 98–99
  Lovell and, 93, 94
  and Milky Way galaxy, 93–94, 95
  Pawsey and, 93, 94
  Ryle and, 93, 94
radio sources. *See also* quasars
  black holes as, 131–32
  distribution of, 94–97, 99
  Lovell and, 112
  Mills and, 96–97
  physical nature of, 112–14, 115
  Reber and, 95, 112
  redshift effect and, 113, 126
  Ryle and, 96–97, 112, 128
  Schmidt and, 112–13, 115, 228
  Slee and, 96–97
radio stars. *See* quasars; radio sources
radio telescope: Reber invents, 94
Reber, Grote: "Cosmic Static," 94, 95
  invents radio telescope, 94
  and radio sources, 95, 112
redshift effect: Hubble measures, 41–43, 78, 88–89, 111, 189
  Humason measures, 41–42, 45, 52, 78, 88, 189
  and radio sources, 113, 126
  Slipher measures, 38–39, 40, 41, 42, 52, 78
Rees, Martin, 128, 142
  and "Big Bang" theory, 173
  and black holes, 135, 173, 225, 226
  on cosmology, 173–74

and gravity waves, 158

and quasars, 132–33

and steady-state theory, 126

"Reflections on Progress, Peaceful Coexistence, and Intellectual Freedom" (Sakharov), 215

"Relation Between Distance and Radial Velocity Among Extragalactic Nebulae, A" (Hubble), 42

relativity, general: basic priincples of, 4–6, 37, 103

    Bekenstein and modifications to, 217–18, 221

    "Big Bang" theory and, 216

    black holes and, 118, 123–24, 131–32, 134, 205, 226

    Bondi and, 98, 119, 154

    and dark matter, 217

    and dialectical materialism, 70–71

    Dicke and, 111–12, 175–76, 212

    difficulty in understanding, x–xi, xii–xiii, 28

    Dirac and modifications to, 213, 217–18, 221

    Dyson on, 81

    Eddington promotes, ix–xi, 18, 22–23, 26, 30, 50, 64, 86

    effect on astronomy, 29

    effects on spacetime, 6, 21, 103–4, 107, 118–20, 124–25, 153, 163, 179, 197, 202–3

    Ellis and, 231

    and evolution of universe, 29–30, 33–35, 36–37

    Feynman and, 109–10, 111

    field equations of. *See* field equations

    galaxies in, 177, 220

    Gödel and, 75–76, 78–80

    Gold on, 117

    Goudsmit and, 81

    gravity in, 6, 9–10, 15–16, 20–22, 29–30, 143–45, 211, 220

    and gravity waves, 168

    Hawking and, 231

    as key scientific concept, xi–xiv, 234

    Lemaître promotes, 36, 71

    Lenard rejects, 68

    limitations of, xiv, 207–8, 217, 224

    neutron stars and, 160–61

    Newtonian physics and, 15, 21, 90–91

    observational evidence for, xii, xiv, 13, 22–26, 36, 37, 39, 50, 52, 62, 68, 78, 80, 111–12, 162, 175–76, 212, 220–21, 224–25, 232–33, 234

    Oppenheimer and, 59, 60–62, 80–81

    and origin of universe, 43–46

    as orthodoxy, 210–11, 212–13

    Peebles on, 216–17

    Penrose and, 118–21, 231

    as "perfect theory," xii, 212

    popular interest in, 25–27, 228, 234

    and predictability, 111

    proposed modifications to, 209–13, 215, 217–20

    quantum physics and, 44, 47–48, 78, 81, 84, 103, 137, 145, 150–51, 194, 197–98, 211, 216–17

    quasars and, 116–17, 126, 134

    radio astronomy and, 234–35

    resistance to, xiv

    Sakharov and modifications to, 215

    and satellite missions, 223–25

    Schild and, 114–15

    in Soviet Union, 70–72, 121–24, 230

    Stark rejects, 69

    and string theory, xiv–xv

    tensor-vector-scalar theory and, 219

    Weinberg and, 226

    Wheeler and, 81, 100–105, 110–11, 130, 208

    Zel'dovich and, 130–31

relativity, numerical, 165–66, 169

relativity, special: basic principles of, 50–51, 56, 138–39

    inertial frames of reference in, 4–6, 9, 10, 14, 15

    quantum physics and, 138–40

    Silberstein and, ix–x

relic radiation: in "Big Bang" theory, 176, 178–79, 182, 188–89

    measurement of, 188–89

Research Institute for Advanced Studies, 107

Riemann, Bernhard: and non-Euclidian geometry, 15, 21, 74

Riess, Adam: wins Nobel Prize, 190

Robertson, H. P., 79

Rosen, Nathan, 154

Rovelli, Carlo: and quantum gravity, 202, 206
Royal Astronomical Society, ix, 18, 25, 56–57, 95, 211, 232
Rubin, Vera, 218
    and Andromeda Galaxy, 182–83
Ruffini, Remo, 134
Russell, Bertrand, 18
    and incompleteness theorem, 77
Rutherford, Ernest, 18
Ryle, Martin: attacks steady-state theory, 95–96, 97–98, 120, 126
    background of, 92–93
    and Hoyle, 95
    and radar research, 92, 94
    and radio astronomy, 93, 94
    and radio sources, 96–97, 112, 128
    wins Nobel Prize, 99, 130

Sachs, Rainer: and "Big Bang" theory, 178
Sagan, Carl, x
Sakharov, Andrei: background of, 213
    and cosmology, 213–14
    and field equations, 214
    and gravity, 214
    and modifications to general relativity, 215, 221
    political opinions, 214–15
    "Reflections on Progress, Peaceful Coexistence, and Intellectual Freedom," 215
    and spacetime, 213–14
    Zel'dovich and, 213, 215
Salam, Abdus, 144
    and electroweak force, 142
    and standard model, 143, 200
Salpeter, Edwin, 131–32
satellite missions: general relativity and, 223–24
Schild, Alfred, 221
    background of, 114–15
    forms Center for Relativity, 114–15, 120
Schmidt, Brian: and cosmological constant, 190
    wins Nobel Prize, 190
Schmidt, Maarten: and radio sources, 112–13, 115, 228

Schrödinger, Erwin, xi
    flees Germany, 69
    and quantum physics, 53, 55, 74, 101, 119, 197
    Space-Time Structure, 119
Schutz, Bernard, 159, 171
Schwartz, John: and string theory, 199, 204–5
Schwarzschild, Karl, 111
    death of, 49
    discovers black holes, 49–50, 64, 75, 131, 132
    Eddington on, 48
    "The Effect of Wind and Air-Density on the Path of a Projectile," 48
    and Einstein, 48–50
    and gravity's effects on spacetime, 48–49, 61–62
    solves field equations, 48–50, 51–52, 58, 61–62, 64, 80, 116, 120–21, 125, 134
    in World War I, 48
"Schwarzschild surface," 49–50, 61–62, 118–19, 120, 131, 146, 225, 227
Schwinger, Julian: and quantum electrodynamics, 140, 142
Sciama, Dennis, 121, 132, 138, 142, 144, 149
    accepts "Big Bang" theory, 126, 127
    background of, 119–20
    and black holes, 150
    and Dirac, 119–20
    and gravity waves, 158
    and Hawking, 127–28, 145
    and quasars, 126, 132–33
    and steady-state theory, 119–20, 126
Seaborg, Glenn: discovers plutonium, 63
Serber, Robert, 60
Silberstein, Ludwik: and special relativity, ix–x
Silk, Joseph: and "Big Bang" theory, 178, 182, 188
    and large-scale structure theory, 187
singularities. *See* "Big Bang" theory; black holes
Sirius B: as white dwarf star, 52, 55
Slee, Bruce: and radio sources, 96–97
Slipher, Vesto: measures redshift effect, 38–39, 40, 41, 42, 52, 78

Smarr, Larry: and black holes, 165, 166
Smolin, Lee: and quantum gravity, 201–2,
    204, 206
  on string theory, 204
Smoot, George: and *Cosmic Background
    Explorer,* 188
  wins Nobel Prize, 188
Snyder, Hartland: and black holes, 47–48, 49,
    64, 120, 123
  and stellar evolution and decay, 61–62, 64,
    75, 81, 101, 116
Sommerfeld, Arnold, 54
Southwest Center for Advanced Studies
    (Dallas), 115
Soviet Union: develops atomic bomb, 72,
    121–23, 213–14
  general relativity in, 70–72, 121–24, 230
  purge of physicists in, 72
  quantum physics in, 72–73
Space-Time Structure (Schrödinger), 119
spacetime: canonical approach to, 197–98, 201,
    203–4, 205
  covariant approach to, 197–98, 200, 202, 205
  curvature of, 14–15, 21, 30–31, 32, 37, 78–79,
    180–81, 182
  effect of gravity on, 48–49, 59, 61–62, 227,
    233
  effects of general relativity on, 6, 10, 21,
    103–4, 107, 118–20, 124–25, 153, 163, 179,
    197, 202–203
  effects of inflationary universe model on,
    182
  Einstein and, xi, 215
  geometry of, 73, 180–81, 200–202, 205–7,
    214, 215, 216, 221, 229–30
  observational evidence of, 233–34
  Sakharov and, 213–14
  in string theory, 205–7
  wave theory of, 153
special theory of relativity. *See* relativity,
    special
Spergel, David: and cosmological constant,
    187
  on cosmology, 174
spin networks: Penrose creates, 202
  and quantum gravity, 202, 203
Square Kilometer Array (SKA), 234–25

Stalin, Joseph: "Dialectical and Historical
    Materialism," 70
  and Soviet atomic bomb project, 122–23
standard model of forces, 145, 197, 199
  and dark matter, 174
  Dirac and, 142
  Efstathiou and, 221–22
  Glashow and, 144
  Hawking and, 194
  Salam and, 143, 200
  and string theory, 198, 200–201
  Weinberg and, 143, 200, 226
Star Trek (TV series), x
Stark, Johannes, 4
  attacks Heisenberg, 69
  rejects general relativity, 69
Starobinsky, Alexei, 148
stars: Eddington and, 50–52, 54–55
  globular clusters, 184
  gravity's effect on, 51, 55–57, 131–32
  neutron stars, 61, 104, 129, 139
  quantum physics and, 50–52, 54–56
  sources of energy in, 50–51, 59–61, 87,
    113–14, 115
  supernovae, 163
  x-ray stars, 132–33, 134, 224–25, 226
stars, evolution and decay of, 49–52, 120,
    123–24, 163. *See also* black holes
  Chandra and, 55–57, 64, 115, 129, 139, 211
  Eddington and, 56–57, 64, 65, 114, 124, 125
  Einstein and, 64–65
  Fowler and, 55–56, 139
  Landau and, 104, 130, 139
  Oppenheimer and, 61–62, 64, 75, 81, 100,
    101, 115, 116, 130
  Snyder and, 61–62, 64, 75, 81, 101, 116
  Wheeler and, 116–17, 125–26, 128, 175
  white dwarfs, 52, 54, 55–58, 64, 87, 129, 139
Stars and Atoms (Eddington), 87
static model of universe, 29, 32. *See also*
    steady-state theory
  de Sitter and, 30, 37
  Einstein promotes, 31, 33–35, 36, 37, 90, 92
  instability of, 45
steady-state theory. *See also* static model of
    universe
  Bondi and, 89–91, 96, 97–98, 119

steady-state theory (*cont.*)
    Born on, 91
    Einstein on, 91
    Gold and, 89–91, 96, 98, 110
    Heisenberg and, 91
    Hoyle promotes, 85–86, 89–92, 96, 97, 98,
        113, 119–20, 126, 230
    Milne on, 91
    Rees and, 126
    Ryle attacks, 95–96, 97–98, 120, 126
    Sciama and, 119–20, 126
Steinhardt, Paul: and cosmological constant,
    187
Stern, Otto, 12
string theory: and "Big Bang" theory, 229–30
    Calabi-Yau geometry and, 199
    Candelas and, 199, 204
    cosmological constant in, 231
    and covariant approach to spacetime, 198,
        202
    DeWitt on, 203
    Duff on, 203–4
    Feynman on, 200
    Friedan on, 200
    Gell-Mann and, 198–99
    general relativity and, xiv–xv
    Glashow on, 200
    gravitons in, 198
    Green and, 199, 204
    Hawking on, 203
    Horowitz and, 199
    hostility toward, 204–5
    M-theory of, 199, 203, 206, 230
    multiverse model in, 231
    principles and flaws of, 198, 199–200
    and quantum gravity, 199, 202, 206
    Schwartz and, 199, 204–5
    Smolin on, 204
    spacetime in, 205–7
    standard model and, 198, 200–1
    Strominger and, 199, 206
    superstrings, 215
    Vafa and, 206
    Witten and, 199, 207, 231
Strominger, Andrew: and string theory, 199,
    206
strong force: in unified theory, 141–42, 143

Sunyaev, Rashid, 188
Supernova Cosmology Project (SCP): and
    cosmological constant, 189–90
supernovae, 216
    and cosmological constant, 189
    and gravity waves, 163
supersymmetry: Hawking and, 194–95

Taylor, Joseph, Jr., 130
    and neutron stars, 160–61, 162, 163
Teller, Edward, 63, 102, 165
tensor-vector-scalar theory of gravity (TeVeS):
    Bekenstein and, 219–20
    and general relativity, 219
Texas Symposiums on Relativistic
    Astrophysics, 118, 121, 128, 135, 161,
    221, 228
    and quasars, 115–17
thermodynamics, second law of. *See* entropy
Thompson, J. J., 18, 25–26
Thorne, Kip, 103, 118
    and black holes, 133–34, 135, 226
    and gravity waves, 167, 169, 171
thought experiments: Bondi performs, 154
    Einstein performs, 3–4, 5–6, 9–10, 22, 71
    Wheeler performs, 103
time travel: Gödel and, 78–79
Tomonaga, Sin-Itiro: and quantum
    electrodynamics, 140
Townes, Charles: wins Nobel Prize, 161
Turner, Herbert, 18–19
Turner, Michael: and cosmological constant,
    187
    on cosmology, 174–75
Tyson, Tony: on gravity waves, 168

Uhuru (satellite), 133
uncertainty principle: in quantum physics,
    53–54
unified theory: CERN and, 142, 144
    Dirac and, 138–39
    Eddington and, 86–87, 137, 212
    Einstein's search for, 73–75, 86, 110, 137,
    211–12
    electromagnetism in, 73, 140–42, 143
    electroweak force in, 142, 143

field equations in, 74
gravity in, 140–41
inflationary universe model in, 180
quantum electrodynamics and, 140–42,
  145
spacetime geometry in, 73
strong force in, 141–42, 143
weak force in, 141–42, 143
United States: Atomic Energy Commission,
  82
Einstein emigrates to, 68–70, 73, 75
Gödel emigrates to, 77
National Aeronautics and Space
  Administration (NASA), 212, 223
universe: CDM model of, 184–85, 186–89
concordance model of, 190–92
de Sitter on age of, 89
distribution of energy and matter in, 28–29,
  32–33, 78
distribution of radio sources in, 94–96, 99
evolution of, 29–30, 33–35, 36–37, 178–79,
  182
expanding model of, 30–31, 33–34, 36–37,
  38–39, 41–46, 48, 71, 88–92, 96, 177
inflation model of, 180–82, 185, 187
origin theories of, 43–46, 89–90, 126
proposed rotation of, 78
static model of, 29–30, 32–35, 36, 45, 85–86,
  90, 92
University of Bern: Einstein as lecturer at, 11
University of Zurich: Einstein as professor at,
  11
Uranus: orbit of, 7

Vafa, Cumrun: and string theory, 206
Volkoff, George, 60–61
Von Neumann, John, 20, 63, 66, 77
Vulcan: as theoretical planet, 8, 217

weak force: gravity as, 218
in unified theory, 141–42, 143
Weber, Joseph: background and personality of,
  155, 170–71
Dyson and, 170–71
experimental methodology questioned,
  157–60

and laser interferometry, 161
searches for gravity waves, 152–53, 154–60,
  163, 170–71, 172
Weber bars: and search for gravity waves,
  155–57, 158–59, 161–63, 171
Weinberg, Steven, 144, 215
and electroweak force, 142
and general relativity, 226
Gravitation and Cosmology, 226, 228
and standard model, 143, 200, 226
Weiss, Rainer, 167
Weyl, Hermann, xi, 19–20, 66, 80, 212
and expanding universe model, 37, 39, 41
Wheeler, John Archibald, 62–63, 120, 133, 144,
  146, 155, 175, 217, 228
background of, 101–2, 108–9
and black holes, 100, 103, 116, 118, 123, 132,
  134, 150, 175, 226
and canonical approach to spacetime, 197,
  201, 203
creates "Wheelerisms," 102–3, 226
and DeWitt, 197
and Feynman, 102
and general relativity, 81, 100–5, 110–11,
  130, 208
and Manhattan Project, 101
and nuclear physics, 100, 104, 122
performs thought experiments, 103
personality of, 101–2
political opinions, 102
and quantum physics, 101, 207
and stellar evolution and decay, 116–17,
  125–26, 128, 175
and wormholes, 103
Wheeler, John Archibald, Charles Misner &
  Kip Thorne: *Gravitation,* 226
Wheeler-DeWitt equation, 197, 201, 202, 203
White, Simon: and cosmological constant,
  186
and galaxy formation, 184
white dwarf stars, 52, 54, 55–58, 64, 87, 129,
  139
Whitehead, Alfred North & Bertrand Russell:
  Principia Mathematica, 18
Wilson, Robert: and "Big Bang" theory, 126–27,
  128, 176, 178
wins Nobel Prize, 176

Witten, Edward, 215
  and string theory, 199, 207, 231
Wittgenstein, Ludwig: and incompleteness
    theorem, 77
Woit, Peter: and quantum gravity, 204
Wolfe, Arthur: and "Big Bang" theory, 178
World War I: British scientists and, 18–19,
    22–23, 35
  Einstein and, 16–17, 19
  Friedmann in, 31–32, 35, 48
  Hubble in, 40
  Lemaître in, 35
  Schwarzschild in, 48
World War II: beginning of, 62
  nuclear physics develops in, 63–64
  radar development in, 87–88, 92–93
World Wide Web, 166
wormholes: Wheeler and, 103

x-ray stars, 132–33, 134, 224–25, 226

Yerkes Observatory (Chicago), 57
Yu, Jer: and evolution of universe, 178–79, 182

Zel'dovich, Yakov: and "Big Bang" theory,
    178, 181, 182, 188
  and black holes, 131–32, 133, 135, 148, 225,
    226, 227
  and cosmological constant, 186, 191, 192
  and general relativity, 130–31
  and large-scale structure theory, 187
  and nuclear physics, 122
  and Sakharov, 213, 215
Zermelo, Ernst, 20
Zwicky, Fritz, 180
  and galaxies, 183